INORGANIC NOMENCLATURE

A PROGRAMMED APPROACH

WESLEY E. LINGREN
Professor of Chemistry
Seattle Pacific University

PRENTICE-HALL, INC., Englewood Cliffs, New Jersey 07632

Library of Congress Cataloging-in-Publication Data

Lingren, Wesley E
   Inorganic nomenclature.

  Bibliography: p.
  Includes index.
  1. Chemistry, Inorganic—Nomenclature—Programmed instruction. I. Title.
QD149.L56   546'.01'4   79-19916
ISBN 0-13-466607-0

Editorial/production supervision
  and interior design: Eleanor Henshaw Hiatt
Cover design: Wanda Lubelska
Manufacturing buyer: Edmund W. Leone

Printed in the United States of America

10 9 8 7 6 5 4 3 2 1

PRENTICE-HALL INTERNATIONAL, INC., *London*
PRENTICE-HALL OF AUSTRALIA PTY. LIMITED, *Sydney*
PRENTICE-HALL OF CANADA, LTD., *Toronto*
PRENTICE-HALL OF INDIA PRIVATE LIMITED, *New Delhi*
PRENTICE-HALL OF JAPAN, INC., *Tokyo*
PRENTICE-HALL OF SOUTHEAST ASIA PTE. LTD., *Singapore*
WHITEHALL BOOKS LIMITED, *Wellington, New Zealand*

CONTENTS

PREFACE

Instructors of high school and college chemistry courses will agree that accurate use of inorganic nomenclature is one of the desired outcomes for their courses. The same goal is true for the authors of many fine chemistry textbooks. Yet, the simple fact is that nomenclature tends to be squeezed out of the modern chemistry curriculum (or at least minimized) because of the heavy pressure of so many other topics. Nomenclature is often found in the appendix or is inserted as a small section in a chapter, and lecturers are hard pressed to find time to amplify it.

This book was written to give students an opportunity to learn an important skill in chemistry and at the same time not cut deeply, if at all, into class time. The approach amplifies the nomenclature sections of most chemistry textbooks and provides many drill exercises. The material in chapters one through five is background and having mastered this material, the student may move to any of the remaining chapters that are appropriate for the material being covered in class.

Many beginning chemistry students seem to feel that the naming of inorganic substances is almost an arbitrary procedure, and, consequently, they often treat the subject in a haphazard way. The purpose of this book is to eliminate the elements of guessing and hoping by providing systematic instruction and practice in naming compounds and writing formulas from the names of commonly encountered chemical substances. The book is not intended to be a handbook of naming for the professional chemist—it is aimed at the undergraduate (or high school) student who is embarking upon a study of chemistry. Upon completion of the book, the student should have sufficient background in the correct use of standard inorganic nomenclature and formula writing for most instructional purposes.

Programmed instruction is particularly well-suited for teaching nomenclature because (1) the subject adapts well to linear programming, (2) a unit, or part of it, can be introduced by the instructor wherever it seems appropriate and without loss of lecture or class time, and (3) the student can learn at his or her own pace.

It is a pleasure to acknowledge the help of the following people and to thank them for their indispensable roles in the preparation of this book: Professors Edward Beardslee and Andrea Norman, Seattle Pacific University, for discussions and advice; Lucia Delamarter for excellent typing and proofreading; my wife, Merrilyn, and children, Eric and Libby, for their continuous encouragement and many sacrifices; and finally, Fred Henry, acquisitions editor; Eleanor Henshaw Hiatt, production editor; and Robert Davis, field representative, of Prentice-Hall for their expert advice and encouragement.

W.E.L.

TO THE STUDENT

This book is designed to bring you to the point where, given a formula for a common inorganic substance, you will be able to name the substance correctly or, given the name, you will be able to write the proper formula, providing you carefully follow the format and instructions given below.

This book is not a textbook in the usual sense of the word, and it is to be used in a different manner. Each chapter consists of: (1) a set of learning objectives, (2) a pre-test on the material in the chapter, (3) a body of programmed subject matter, and (4) a post-test on the chapter.

To use the learning method presented in this book, proceed in the following manner (also summarized in the Flow Diagram on page viii):

1. Examine the learning objectives for the chapter and if they seem to you to be previously learned material, then take the pre-test on the page that follows. The answers to the pre-tests are in Appendix I. Grade yourself, and if you got 90 percent or more, then you may move directly to the next chapter, where you repeat the examination of learning objectives, and so forth.

2. If the learning objectives for the chapter are new or if you received less than 90 percent on the pre-test, then you should move directly into the programmed material of the chapter.

3. Use the programmed material as follows:

   a. Cover everything except the lead paragraph (T–1) and the first question (Q1) with a sheet of paper.

b. Study the lead paragraph; then read the first question (Q1) and write your answer in the space provided or on a sheet of paper.

c. Move the cover paper until you uncover A1 and check your answer against the correct answer. If you were right, move to Q2 and repeat these steps.

d. If you wrote an incorrect answer, return to the lead paragraph T–1 and try to discover where you went astray. It may also be helpful at this point to look up the pertinent chapter in your chemistry textbook or to see the references given at the end of your textbook to aid you further in solving your difficulty.

e. If you are satisfied that you have pinpointed the difficulty with your incorrect response, move to Q2 and write the answer to it. Compare your answer with A2 and repeat for Q3, A3, and so forth.

4. When you complete the programmed material, take the post-test at the end of the chapter. Grade yourself by comparing your answers with the key provided in Appendix II. If you get 90 percent or better, proceed to the next chapter. A score of less than 90 percent indicates that you had better repeat the programmed material, emphasizing those areas where you scored poorly on the post-test.

Final Note: This material is carefully sequenced, and you cannot expect to gain full value from it if you do not follow the steps carefully. If you encounter a section that you believe you know, you can spend less or no time on it. But be sure you read the lead paragraph carefully before skipping the section, because there may be something new in it.

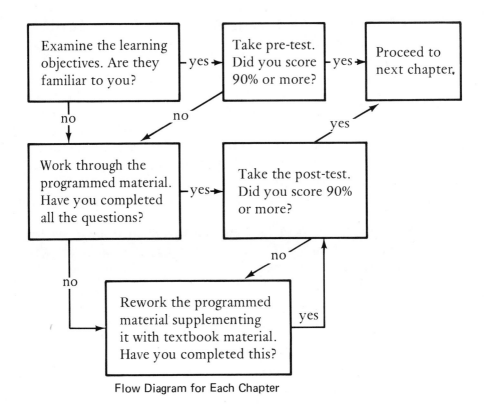

Flow Diagram for Each Chapter

1

PERIODIC TABLE

OBJECTIVES

The student should be able to

1. Express, with particularity, the major attributes, the organizing principle, and the benefits to the chemist of the periodic table.

2. Distinguish between, in terms of chemical properties, a row and a column of the periodic table.

3. Sketch from memory the relative positions of the following in the periodic table: the element hydrogen, the metals and nonmetals, the alkali and alkaline earth metals, the halogens, chalcogens (oxygen family), noble gases, transition metals, lanthanides (rare earths), and actinides.

4. Recognize and use the names and symbols for: the first, second, and third short rows of elements, the first long row, including the first set of transition metals, the family members of the alkali metals, the alkaline earths, the chalcogens, and the halogens.

PRE-TEST

1. (1 pt)   What type of substance is classified in the periodic table?
2. (2 pt)   State the organizing principle of the periodic table.
3. (2 pt)   Give the major benefit that a present-day chemist derives from the periodic table.
4. (1 pt ea)   Where in the periodic table do you expect to find each of the following? (Show by a rough sketch.)
   (a) the halogens
   (b) the metals
   (c) the alkaline earth metals
   (d) the rare earth metals
   (e) the actinides
   (f) the transition metals
   (g) hydrogen
5. (½ pt ea)   Give the names and symbols for each of the following.
   (a) four members of the chalcogens
   (b) six members of the first transition metals series
   (c) the two members of the shortest row
   (d) two members of the alkali metals
   (e) two members of the alkaline earth metals
6. (2 pt)   In terms of chemical properties, how does a row differ from a column in the periodic table?
7. (½ pt ea)   For each of the listed elements, state the family name.
   (a) fluorine
   (b) magnesium
   (c) sulfur
   (d) sodium

INTRODUCTION

The purpose of this chapter is to provide you with an introduction to or review of the organization of the chemical elements in the periodic table. This material is found in a book on nomenclature because, provided that you know and can use the family relationships between elements, you can write many formulas and name many compounds by analogy. The demand on your memory will thus be reduced. For example, knowing that bromine and iodine belong to the same chemical family as chlorine and that HCl is named hydrogen chloride you could predict that HBr and HI are likely compounds and that their names would be hydrogen bromide and hydrogen iodide, respectively. Many other cases of this type can be found.

The chemical elements can be identified by their chemical and physical properties. Those traits of an element that refer to its ability to react with other substances are called **chemical properties**. The observation that sodium metal reacts spontaneously with water to form hydrogen gas and sodium hydroxide or the fact that helium gas will not react with other substances are statements of chemical properties for these two elements.

The **physical properties** of a substance are those traits that can be measured without changing the composition of the substance. Color, density, and normal melting point are examples of physical properties.

1-T-1

When the properties of the 103 chemical elements are surveyed with the intent to group elements that show similar sets of chemical traits, it quickly becomes apparent that several different groups can be found. Refining the classification scheme by including the physical properties, particularly the number of protons or the atomic number of the element, leads to the grouping of the elements found in the present-day periodic table (Table 1-1). The organizing principle of the table is summarized in the **periodic law**, which states that when the elements are arranged in order of increasing atomic number, similarities in chemical and physical properties occur at regular intervals.

Table 1-1. The Long Form of the Periodic Table

1A	2A	3B	4B	5B	6B	7B		8B		1B	2B	3A	4A	5A	6A	7A	8A
																H	He
Li	Be											B	C	N	O	F	Ne
Na	Mg											Al	Si	P	S	Cl	Ar
K	Ca	Sc	Ti	V	Cr	Mn	Fe	Co	Ni	Cu	Zn	Ga	Ge	As	Se	Br	Kr
Rb	Sr	Y	Zr	Nb	Mo	Tc	Ru	Rh	Pd	Ag	Cd	In	Sn	Sb	Te	I	Xe
Cs	Ba	La*	Hf	Ta	W	Re	Os	Ir	Pt	Au	Hg	Tl	Pb	Bi	Po	At	Rn
Fr	Ra	Ac†															

* Lanthanides.

† Actinides.

Q1. The periodic table is an organization of the 103 chemical _____ based upon their properties.

A1. elements

Q2. The periodic table is an arrangement of the chemical elements based upon the _____ number and _____ in chemical properties.

A2. atomic, similarities

Q3. If you arrange the chemical elements in order of increasing atomic numbers, you will find that similarities in _____ properties recur.

A3. chemical/physical

Q4. The periodic law states that as you increase the atomic number, _____ and _____ properties recur.

A4. physical, chemical

Q5. One of the most obivous features of the periodic table is the _____ of chemical properties.

A5. recurrence

1-T-2

A very common form of the periodic table is given in Table 1–1. You can quickly discover that there are seven horizontal rows (don't overlook the very short first row). The lanthanides and actinides are inserted at the asterisks and are not considered separate rows. You may already know that the rows are called **periods** or **series**. In general, as you move from left to right across a row, the properties of the elements change gradually and regularly from highly metallic on the left to distinctly nonmetallic on the right.

Q1. A period is a _____ in the periodic table.

A1. row

Q2. A row is called a period or a

_____ .

A2. series

Q3. The shortest period has _____ elements in it.

A3. two

Q4. As one moves from left to right in a period the chemical and physical properties tend to change

_____ .

A4. gradually

Q5. You would *not* expect elements that are beside each other in a period to show identical _____ and _____ properties.

A5. chemical, physical

Q6. Elements that lie at the far right of the periodic table are metallic/non-metallic?

A6. nonmetallic

Q7. Cesium is found at the lower left corner of the periodic table. You'll predict that it will be a metal/non-metal?

A7. metal

1-T-3

Referring again to Table 1–1, you will see that there are 16 vertical columns headed by a number–letter combination (note that three columns are lumped together and are counted as one under heading 8B, and further, the lanthanides and actinides are *not* counted as separate columns). The 16 columns are called **families** or **groups**. Within the members of a family the chemical properties closely resemble one another.

Q1. A period is a row and a family is a _____ in the periodic table.

A1. column

Q2. The primary feature held in common by the members of a chemical family is that their properties are

_____ .

A2. similar

Q3. Although potassium and calcium are next to each other and are both metals, their chemical properties are not alike; therefore, they are not members of the same _____.

A3. family

Q4. Elements X and Y resemble each other in their chemical reactions, so they are likely to be members of the same period/series/family.

A4. family

1-T-4

The families listed under A's are called **representative elements** and those with B's are called **transition metals**. Furthermore, chemists use special names for certain families. You should memorize the names of these families, their members' names and symbols, and their location on the periodic chart. The important ones are 1A, alkali metals; 2A, alkaline earth metals; 6A, chalcogens or oxygen family; 7A, halogens; and 8A, noble gases (formerly inert gases). At this point you should simply memorize the member's names and symbols for each of these families and each group's location in the periodic table. After you have done this, proceed with the following questions.

Q1. Where, relative to the alkali metals, will you find the alkaline earth metals?

A1. next column to the right

Q2. List the names and symbols of the chalcogens.

A2. oxygen, O
sulfur, S
selenium, Se
tellurium, Te
polonium, Po

Q3. Give the symbols of the halogens and the family location in the periodic table.

A3. F, Cl, Br, I, At; column 7A is second from the right-hand edge

Q4. Complete the following sentence: Group 1A is called the _____ metals and consists of (spell the names correctly) lithium, _____, _____ , _____ , _____ and francium.

A4. alkali, sodium, potassium, rubidium, cesium

Q5. What is the name and symbol of the heaviest noble gas?

A5. radon (Rn)

Q6. Helium belongs to the halogen family. true/false

A6. false

Q7. Helium, neon, and krypton all belong to the _____ family.

A7. inert or noble gas

Q8. The chalcogens are located just to the left of the halogens. true/false

A8. true

Q9. The last column on the right consists of helium and other members of the noble gases. true/false

A9. true

1-T-5

To summarize, a row or period is characterized by a gradual change in chemical and physical properties as you move across it, whereas a column or family is identified by the very close resemblance of the chemical properties of its members as you move down (or up) the column and, to a lesser extent, their similar physical properties. Special names are often used when referring to certain chemical families.

Q1. Although the chemical properties differ in a ———————, they resemble one another in a

————————.

A1. period, family

Q2. To distinguish between a period and a family you would describe the different trends in the respective members' ——————.

A2. properties

Q3. Magnesium and calcium closely resemble one another in physical and chemical traits; therefore, you would not expect to find them in the same —————— in the periodic table.

A3. row (period or series)

Although the periods have no special names like some of the families, it is still very useful to memorize the names and symbols in order for the following rows: the shortest row (two members); the second, third, and fourth rows, including the first set of transition metals (Sc through Zn). The memory work asked for is unexciting. Few will argue with that; however, you must try to appreciate that these names and symbols are as vital to chemistry as nouns and verbs are to another language. Just as you must memorize some basic vocabulary in a foreign language, so it is here. Take some time and be serious and systematic about the memory work and you will be well repaid during your study of chemistry.

Q1. Where will you find the lightest element, hydrogen?

A1. first row, column 7A; some tables show it first row, column 1A; it has some properties of both groups

Q2. What are the names and symbols of the elements in the first eight-member row?

A2. lithium, Li
beryllium, Be
boron, B
carbon, C
nitrogen, N
oxygen, O
fluorine, F
neon, Ne

Q3. Give the symbols for the first set of transition elements.

A3. Sc, Ti, V, Cr, Mn, Fe, Co, Ni, Cu, Zn

Q4. Phosphorus is in the first, second, or third period?

A4. third

Q5. The symbol for silicon is

_____ .

A5. Si

Q6. He belongs to the _____
period.

A6. first

Q7. Rearrange the following into the order
of increasing atomic numbers: O, B,
F, Na, Be, C.

A7. Be, B, C, O, F, Na

1-T-7

The similarities of the chemical traits within a family form the basis for
the most useful property of the periodic table, because if you know the
members of the family and the chemistry of a typical member of the
family, you can infer the chemistry for the other elements in that
group. This extension can also apply to formulas of analogous compounds within the family and therefore to the names of the compound.
For example, if you know that sodium reacts spontaneously with water
to form hydrogen gas and sodium hydroxide, NaOH, you can infer that
any one of the alkali metals will react spontaneously with water to
form hydrogen gas and the corresponding metal hydroxide, MOH.

For nomenclature purposes you do not need all the specific reaction details, but you do need the formula and name of the representative product if you are to draw the correct inference.

Q1. The hydroxide of magnesium has the
formula, $Mg(OH)_2$. What is the formula for calcium hydroxide?

A1. $Ca(OH)_2$

Q2. If NaH is called sodium hydride,
what would you predict for the name
of LiH?

A2. lithium hydride

Q3. Since Na, Mg, and Al all fall into the same period, it is reasonable to expect that they all have oxides with the formula MO_2 (M = metal). true/false

A3. false (inference of like formula is true for family but not for period)

Q4. Aluminum, boron, and gallium are members of group 3A. What formula would you expect for the chloride of aluminum if Ga and B exhibit formulas $GaBr_3$ and BF_3, respectively?

A4. $AlCl_3$

Q5. A common acid of sulfur is H_2SO_4, sulfuric acid. Give the formula of the analogous acid for selenium.

A5. H_2SeO_4

SUMMARY

The periodic table is an arrangement of the elements based on their atomic numbers. Similarities in physical and chemical properties occur at regular intervals. A row of elements is called a period and is characterized by a gradual change from metallic to nonmetallic properties as you move across it from left to right. Elements located in vertical columns are called families and are characterized by close similarities in chemical properties. Families of elements often have such special names as halogens, alkali metals, and the noble gases. Knowledge of the properties of a family member can often be used to infer the behavior of other members of the group.

1. (2 pt)   Distinguish between a period and a family in the periodic table.

2. (½ pt ea)   Give the family name for each listed element.

    (a) Ne

    (b) Cl

    (c) Mg

    (d) S

3. (½ pt ea)   List the names and symbols for members of the alkaline earth family.

4. (½ pt ea)   Circle those elements that belong to the first row of transition metals: Na, Ca, Cr, Zn, Al, Li, Ti, Sr, Cs.

5. (2 pt)   What is the organizing principle of the periodic table?

6. (2 pt)   State the major benefit derived by the chemist from the periodic table.

7. (1 pt ea)   Sketch a rough version of the periodic table (long form) and show where to find each of the following:

    (a) halogens

    (b) noble gases

    (c) transition metals

    (d) shortest row

    (e) alkali metals

2

ELECTRONEGATIVITY

OBJECTIVES

The student should be able to

1. Define, with an example, the electronegativity of an element.
2. State the trend in electronegativity in both a period and a family of the periodic table.
3. Differentiate between electronegative and electropositive elements.

1. (1 pt)   What is electronegativity?
2. (2 pt)   State the trend in electronegativity as you move
             (a) left to right across a period
             (b) from bottom to top in a chemical family
3. (1 pt ea) Circle the more electronegative element in each of the following pairs.
             (a) F or Cl
             (b) O or F
             (c) Na or Li
             (d) Li or N
             (e) H or C
4. (1 pt)   Where is the most electronegative element located on the periodic chart?
5. (1 pt)   Where is the most electropositive element?

INTRODUCTION

The characteristic of an element called electronegativity plays an important role in the naming process because it generally determines which element is placed first in a binary molecular formula and name.

It is assumed throughout the remaining chapters of this guide that you have mastered the objectives of Chapter 1. This means that you have in your memory important facts regarding the location of important chemical families in the periodic table, certain trends and characteristics seen in the table, and the names and symbols of about 40 common elements. If you are rusty or do not know them, you should backtrack and spend some time getting this material firmly into your mind before you move ahead.

2-T-1

Elements or atoms exhibit a tendency to attract shared electrons, which is called **electronegativity** (EN). Initially, Linus Pauling arbitrarily assigned electronegativity values in the following way:

Li	Be	B	C	N	O	F
1.0	1.5	2.0	2.5	3.0	3.5	4.0

A moment of inspection reveals that these values increase by one-half unit as you move left to right across the period. Numerical values of electronegativity, which depend upon such factors as bond energy and dipole moment, have been calculated relative to Pauling's values for nearly all the elements. A partial list is given in Table 2–1.

Q1. The property of an atom that is a measure of its ability to attract shared electrons is _____ .

A1. electronegativity

Q2. If an atom did not show any attraction for shared electrons, its electronegativity would be _____ .

A2. zero

Table 2-1. Abbreviated Table of Electronegativities

H															
2.1															
Li	Be										B	C	N	O	F
1.0	1.5										2.0	2.5	3.0	3.5	4.0
Na	Mg										Al	Si	N	S	Cl
0.9	1.2										1.5	1.8	2.1	2.5	3.0
K	Ca	Sc	Tl	V	Cr	Mn	Fe	Co	Cu	Zn	Ga	Ge	As	Se	Br
0.8	1.0	1.3	1.5	1.6	1.6	1.5	1.8	1.8	1.9	1.6	1.6	1.8	2.0	2.4	2.8
Rb															I
0.8															2.5
Cs															At
0.7															2.2

Q3. If atom A has a smaller electronegativity than atom B, you could say that A has a greater/smaller tendency to attract shared electrons than B.

A3. smaller

Q4. Note that atom F has the largest electronegativity of all the atoms. Would you expect to find any other atom that could pull shared electrons away from F?

A4. no

Q5. If electropositive is the opposite of electronegative, would the least electronegative element be the most electropositive?

A5. yes

Q6. Is it true that the most electropositive element has the smallest tendency to attract shared electrons?

A6. yes

2-T-2

For the purposes of nomenclature we need only to know certain trends and the relative EN positions of the more frequently encountered elements. Examine Table 2–1 carefully, noting the trends in EN as you go across a period and again as you move up (or down) a family (these trends are paralleled by the missing rows).

Q1. As you move across a period from left to right the value of EN increases/decreases.

A1. increases

Q2. Without looking at Table 2–1, Li is more/less electronegative than F.

A2. less

Q3. If the order of elements from left to right in a period is Sc, Ti, Cr, Cu, Zn, you would predict that _____ is the more electropositive.

A3. Sc

Q4. The most/least electronegative element in a family is found at the bottom of the column.

A4. least

Q5. Magnesium lies above radium in the alkaline earth metal family. Would you expect Mg to be more/less electronegative than Ra?

A5. more

Q6. Suppose that elements, A, B, C, and D are members of a chemical family and that they are listed in order of increasing atomic number. Which would be the most electropositive?

A6. D

Q7. In general, are nonmetals more/less electronegative than metals?

A7. more

Q8. Where on the periodic table is the most electronegative element?

A8. upper right-hand corner

Q9. If you drew arrows pointing in the direction of increasing EN for periods, they would have their heads on the left/right.

A9. right

Q10. If you drew arrows pointing in the direction of increasing EN for families, they would point up/down.

A10. up

Q11. Suppose that an arrow is drawn on the periodic table connecting the least electronegative with the most electronegative element. To which corner does it point?

A11. upper right

Q12. In general, does the trend of increasing EN flow from lower left to upper right in the periodic table?

A12. yes

SUMMARY

You do *not* need the numerical values for EN; however, you should know the trends for EN in the periodic table. A careful look at Table 2-1 will reveal that as you move across a period from left to right the value of EN tends to increase. You will also find that moving upward from the bottom of a family shows an increase in EN. The least electronegative element is located in the bottom left corner of the chart and the most electronegative is located in the upper right corner (excluding the noble gases). You can also discover that metals tend to be much less electronegative than do nonmetals (compare O, F, N, Cl with typical metals). The opposite of electronegative is electropositive; for example, chemists frequently say that a metal is more electropositive than a nonmetal. This is simply another way of saying that the metal is less electronegative than the nonmetal.

Q1. The general trend in EN going left to right across any given period is a gradual _____.

A1. increase

Q2. The most electronegative element is

_____ .

A2. fluorine

Q3. Which of the family members oxygen, sulfur, and selenium would be most electropositive?

A3. selenium

Q4. Where in the periodic table would you expect to find the most electropositive element?

A4. lower left corner

Q5. Carbon and sulfur exhibit EN values that are approximately the same. If a covalent bond is formed between the two elements, would you expect that the bond electrons are likely to be closer to carbon?

A5. no; they will be approximately equidistant between the two

Q6. Arrange the following members of the second row in order of increasing EN: P, Si, Al, Mg, Cl.

A6. Mg, Al, Si, P, Cl

Q7. If you move up a periodic table family, you would expect EN to _____ .

A7. increase

Q8. Generally speaking, metals are more electropositive than are nonmetals. true/false

A8. true

POST-TEST

1. (1 pt)   Write a brief qualitative definition of electronegativity.
2. (1 pt ea)   Circle the most electropositive element of each pair.

    (a) Li and Na

    (b) Li and Mg

    (c) Na and F

    (d) C and H

    (e) Al and Si

3. (1 pt)   What element is the most electronegative and where is it located in the periodic table?
4. (1 pt)   How does EN change as you move up a periodic family?
5. (1 pt)   What is the trend in EN as you move right to left across a period?
6. (1 pt)   Moving from bottom left to upper right on the periodic table, how does EN vary?

CHAPTER

3

OXIDATION NUMBER AND VALENCE

OBJECTIVES

The student should be able to

1. Write a definition of oxidation number and valence.
2. Distinguish between oxidation number and valence.
3. Find the oxidation number of an element given the formula of a compound or ion containing the element.
4. Compare oxidation number and oxidation state.
5. Give from memory the most common oxidation number(s) for
   (a) alkali metals
   (b) alkaline earth metals
   (c) oxygen
   (d) hydrogen

1. (1 pt)   What is the most common oxidation number for hydrogen in its compounds?

2. (1 pt)   What is the oxidation number for the alkaline earth metals?

               What would be the oxidation state for these elements?

3. (2 pt)   State the oxidation number of sulfur in $H_2S_2O_3$.

4. (1 pt)   Give the valence of carbon in CH_4.

5. (2 pt)   Distinguish between the valence of carbon and its oxidation number in $C_6H_{12}O_6$.

6. (1 pt)   If chlorine has an oxidation number of 1-, what is the combining power of a chlorine atom in $AlCl_3$?

7. (1 pt) Na_2O_2 is the formula of sodium peroxide. What is the oxidation number of oxygen in this compound?

8. (1 pt)   When hydrogen combines with an element less electronegative than itself, it forms an hydride. What is the oxidation number for hydrogen in such compounds?

INTRODUCTION

Essential to an understanding of the material presented in this chapter is the recollection that an ion is an electrically charged particle that can be either simple or complex. Mononuclear ions such as Na^+ and Cl^- are typical simple ions, while more complex, many-nuclear ions can be represented by SO_4^{2-} and $Cr(H_2O)_6^{3+}$. Positively charged ions are called **cations** and negatively charged ions are called **anions**.

It is also essential to remember that a molecule is an electrically neutral species composed of two or more atoms chemically bonded together. Although most molecules consist of different elements bonded together, such as $C_{12}H_{22}O_{11}$ and NO_2, there are several molecules that are composed of only one element, for example, H_2 and P_4.

In discussing and using the nomenclature and formulas of molecules and ions, frequent reference is made to oxidation numbers, oxidation state, and to a lesser extent, valence.

3-T-1

The **valence** of an element is the combining capacity of the element, or to say it in slightly expanded way, it is a measure of the capacity of the element to form chemical bonds. The valence is always a positive whole number.

> Q1. The combining capacity of an element is its _____.

A1. valence

> Q2. When using the word "valence" properly, one will be stating a property of an _____.

A2. element

> Q3. When magnesium forms a compound, we generally find it participating in two chemical bonds; therefore, the valence of Mg must be _____.

A3. two

Q4. The structural formula for carbon

$$Cl$$
$$|$$
tetrachloride is $Cl-C-Cl$, where each
$$|$$
$$Cl$$

line represents a chemical bond. What is the valence of carbon in this molecule?

A4. four

Q5. What is the combining capacity of chlorine in CCl_4?

A5. one

Q6. A student states that the valence of the oxide ion, O^{2-}, is "minus two." Is he, strictly speaking, correct?

A6. no; valence is always positive

Q7. What did the student actually state about the ion in Q6?

A7. its electrical charge

Q8. What is the valence of oxygen in $(Na^+)_2 (O^{2-})$?

A8. two

Q9. The answer to Q8 is two because the oxygen participates in _____ chemical bonds.

A9. two

Q10. The valence of an element is a measure of the _____ of the element to form chemical bonds. It is always a _____ integer.

A10. capacity, positive

3-T-2

The valence of an element is not always fixed; in fact, about 80 percent of the elements have variable valences. A typical example is iron, which shows combining capacities of two and three in $FeCl_2$ and $FeCl_3$, respectively.

Q1. Most elements have _____ valences.

A1. variable

Q2. An element with a variable valence has more than _____ combining capacity.

A2. one

Q3. Two compounds of copper are $CuBr_2$ and $CuBr$. Copper shows valences of _____ and _____ in these compounds.

A3. two, one

Q4. Assume that the valences of gold are one and three. What would be the formula of the compounds of gold and an element X that shows a valence of one?

A4. AuX and AuX$_3$

Q5. As a general rule it is not surprising to find that an element has _____ valence.

A5. variable (more than one)

3-T-3

Another term that is usually encountered more frequently than valence in a discussion of nomenclature and formulas is oxidation number. The **oxidation number** is defined as a positive or negative number that represents the charge that an atom appears to have in a given species when the bonding electrons are counted using some rather arbitrary rules. Comparing the definition of an oxidation number with that for valence, you should note that they are not, strictly speaking, the same thing. Valence refers to the ability to combine with other elements and it is always positive, whereas the oxidation number is an apparent electrical charge and may be either positive or negative.

Q1. An oxidation number is an apparent _____ of an atom in a chemical species.

A1. charge (electrical charge)

Q2. The sulfur atom in SO_4^{2-} appears to have an electrical charge of 6+ units; the oxidation number of S is _____ .

A2. 6+

Q3. The oxidation number of oxygen in OH$^-$ is 2–. What is the apparent charge of the oxygen atom in this species?

A3. 2–

Q4. The Cl^- ion has a valence of 1. What is its oxidation number?

A4. 1–

Q5. If you determine the apparent charge on an atom in a species, you have found the atom's _____

A5. oxidation number

Q6. Valence is always positive, whereas an oxidation number may be

_____ .

A6. positive, negative, or zero

Q7. Valence is a measure of an atom's _____ and the oxidation number is its _____ .

A7. combining capacity, apparent electrical charge

3-T-4

Oxidation numbers for the various atoms can be assigned using the following set of rules:

Rule 1.* In free elements, each atom always has an oxidation number of zero.

Q1. Elemental fluorine exists as F_2. What is the oxidation number for each F atom in fluorine gas?

A1. 0

*These rules are adapted from Michell J. Sienko and Robert A. Plane, *Chemistry: Principles and Properties* (New York: McGraw-Hill Book Company, 1966), pp. 85–88.

Q2. Elemental sodium metal consists of a three-dimensional array of sodium atoms. The representative formula for the metal is Na. What is the oxidation number for each sodium atom in this crystal?

A2. 0

Q3. The oxidation number for elemental sulfur, which is found in the molecule S_8, is zero. This means that we assign an apparent charge of _____ to each sulfur atom in S_8.

A3. 0

Q4. The element phosphorus exists as P_4 molecules. What is the proper oxidation number for P atoms in this case?

A4. 0

3-T-5

Rule 2. In compounds containing the alkali metals, the oxidation number of the metal is 1+. In compounds containing the alkaline earth metals, the oxidation number of the metal is 2+.

Rule 3. In compounds containing oxygen, the oxygen atom is generally given an oxidation number of 2−. Two exceptions: (a) the peroxide ion O_2^{2-} has two oxygen atoms sharing a total charge of 2−, thus each oxygen must have an oxidation number of 1−; (b) in a rare instance O may have a value of 2+, for example, in OF_2. This unusual case occurs when O is bound to an element of higher electronegativity. The very electronegative F atoms would draw the bonding electrons away from the oxygen nucleus, thereby leaving it deficient in electrons.

Q1. Which is the more common oxidation number for oxygen: 2−, 1−, or 2+?

A1. 2−

Q2. What is the expected oxidation number of each atom in MgO? Mg = _____ and O = _____.

A2. 2+, 2−

Q3. What is the oxidation number of oxygen in Na_2O?

A3. 2−

Q4. What is the oxidation number of oxygen in Na_2O_2?

A4. 1−

Q5. Aluminum oxide has the formula Al_2O_3. What is the oxidation number of oxygen in this formula?

A5. 2−

Q6. What is the oxidation number of oxygen in BaO_2?

A6. 1−

3-T-6

Rule 4. In compounds containing hydrogen, the oxidation number of hydrogen is generally 1+. One exception: When hydrogen is bonded to elements less electronegative than itself, it is assigned an oxidation number of 1−; thus in LiH the oxidation number of hydrogen is 1−.

Q1. The most common oxidation number for hydrogen is ———————— .

A1. 1+

Q2. The oxidation number of hydrogen in HCl is ———————— .

A2. 1+

Q3. You expect an oxidation number for hydrogen of ————————— when it is bonded to an element less electronegative than itself.

A3. 1–

Q4. What is the oxidation number for H in MgH_2?

A4. 1–

3-T-7

Rule 5. In any molecule or ion, the sum of the positive and negative oxidation numbers must equal the net charge on the species.

For example, consider the species Na_2O; we know from Rule 2 that each Na ion contributes 1+, for a total apparent charge of 2+. Further, Rule 3 assigns the oxygen atom a 2– oxidation number, which is also the total apparent charge. The sum $[2 \times (1+)] + [1 \times (2-)] = 0$, which is the proper net value for a neutral species.

As another example, consider the hydroxide ion, OH^-. Following the rules again, one hydrogen contributes 1+ and one oxygen is assigned 2–. The sum $[1 \times (1+)] + [1 \times (2-)] = 1-$, which is the net charge on this ion.

Q1. Assume that the formula for potassium oxide is KO. Is this consistent with the rules?

A1. no

Q2. What would be the correct formula for potassium oxide?

A2. K_2O

Q3. Which of the following would be a correct formula for magnesium chloride: $MgCl^{1+}$, $MgCl_2$, or Mg_2Cl^{2+}?

A3. $MgCl_2$

3-T-8

Growing directly out of Rule 5 is a simple process for determining the oxidation number for an element not treated by the other rules. To illustrate, what is the oxidation number of chlorine in $HClO_4$? Remember that the sum of the positive and negative oxidation numbers must equal zero for this neutral molecule; therefore, write

$$[1 \times (1+)] + [1 \times (?)] + [4 \times (2-)] = 0$$

$$\text{for H} \qquad \text{for Cl} \qquad \text{for O}$$

Manipulating this equation so that (?) is isolated, we get (?) = 8 − 1 = 7+, which is the missing value for the oxidation number for chlorine in $HClO_4$.

Another case: What is the oxidation number of chlorine in the chlorite ion, ClO_3^-? Following the same procedure as shown in the first example, we write

$$[1 \times (?)] + [3 \times (2-)] = 1-$$

$$\text{for Cl} \qquad \text{for O}$$

Note that in this case the left side is set equal to 1-, the net charge on the ion. Solving for (?), we get (?) = 6 - 1 = 5+. In this case chlorine has an oxidation number of 5+.

Q1. What is the oxidation number of chlorine in NaCl?

A1. 1-

Q2. What is the oxidation number of selenium in SeO_2?

A2. 4+

Q3. What is the oxidation number of sulfur in H_2SO_4?

A3. 6+

Q4. What is the oxidation number of aluminum in Al_2O_3?

A4. 3+

Q5. What is the oxidation number of Pt in $PtCl_6^{2-}$ assuming that the oxidation number of Cl is 1-?

A5. 4+

Q6. If Cl and OH both have oxidation numbers of 1-, what is the oxidation number of cobalt in $[Co(OH)_3(Cl)_3]^{3-}$?

A6. 3+

Q7. What is the oxidation number of each chromium atom in $K_2Cr_2O_7$?

A7. 6+

Q8. What is the oxidation number of nitrogen in Mg_3N_2?

A8. 3−

Q9. What is the oxidation number of nitrogen in NH_3?

A9. 3−

Q10. What is the oxidation number of nitrogen in N_2O_3?

A10. 3+

3-T-9

Because the rules for determining oxidation numbers are based upon an arbitrary way of counting electrons, you will occasionally run into a weird-looking oxidation number. For example, the oxidation number of iron in Fe_3O_4, found by following the rules, is 8/3+. If you try to correlate this with a whole number of electrons, which would not be unreasonable since the 8/3+ represents an atom's charge, you will be frustrated. Fe_3O_4 is actually a 1:1 mixture of FeO and Fe_2O_3. In cases like this, you should recall the definition of oxidation number, which states that it is the *apparent* charge on the atom, not the *actual* charge. For the purposes of balancing equations and writing formulas it is not necessary to know the actual charge; the apparent charge is perfectly acceptable as long as we follow the rules governing its determination and use.

Q1. What is the oxidation number of each carbon atom in $C_6H_{12}O_6$?

A1. zero

Q2. Does this mean that there is actually no electrical charge involved in the bonding of carbon in this molecule?

A2. no

3-T-10

The term "oxidation number" is frequently used almost synonymously with **oxidation state**; thus chlorine, with an oxidation number 7+, is also said to be in the 7+ oxidation state.

> **Q1.** If the oxidation number for barium is 2+, its oxidation state must be
>
> _____ .

A1. 2+

> **Q2.** What is the oxidation state of bromine in the BrO_3^- ion?

A2. 5+

SUMMARY

The oxidation number can be determined for an atom following a set of rules. This number is sometimes referred to as the oxidation state. It is not always the actual charge on the atom nor is it always the valence, although sometimes it coincides with both. We will use it in naming compounds and ions and it can also be used in balancing oxidation–reduction equations.

POST-TEST

1. (1 pt)   What is the common oxidation number for the alkali metals?
2. (2 pt)   Distinguish between oxidation number and valence.
3. (1 pt)   What is the oxidation number of carbon in C_2H_6O?
4. (1 pt)   If fluorine has an oxidation number of 1–, what is its valence in MgF_2?
5. (1 pt)   What is the oxidation number of platinum in PtI_4^{2-}?
6. (1 pt)   The following is called potassium hydride, KH. What is the oxidation number of hydrogen in this compound?
7. (1 pt)   Give the oxidation number of chromium in $K_2Cr_2O_7$.
8. (2 pt) OF_2 is a compound in which oxygen fails to exhibit its usual oxidation number. What is the oxidation number in OF_2, and what would you normally expect it to be?

ELEMENTS

OBJECTIVES

The student should be able to

1. Recognize and state the names and symbols from memory for the first 40 elements in the periodic table.

2. Recognize and use properly the alternative Latin names for the following common elements:
   (a) copper
   (b) gold
   (c) iron
   (d) lead
   (e) silver
   (f) tin

3. Position correctly around the atomic symbol each of the following:
   (a) mass number
   (b) atomic number
   (c) number of atoms
   (d) ionic charge

1. (3 pt)   Give the atomic symbol for each of the following: titanium, sodium, argon, fluorine, chromium, selenium, and lead.

2. (5 pt)   Write the correct trivial name and Latin name, where appropriate, for each of the following symbols: Cu, Sn, Cl, Li, Ag, W.

3. (2 pt)   Write the complete meaning of the representation $^{18}_{8}O^{2-}_{2}$.

INTRODUCTION

In 1970, culminating many years of work by many people, a set of definitive rules for naming elements, ions, compounds, and other chemical species was published in the report of the Commission on the Nomenclature of Inorganic Chemistry of the International Union of Pure and Applied Chemistry (IUPAC). The report was an important forward step in the development of a comprehensive and systematic inorganic nomenclature which is to be used throughout the world.

Committees from the American Chemical Society and the National Research Council reviewed the IUPAC report and published an American version of the rules, which does not differ substantially from those of the IUPAC commission. The rules given in this book are taken from the American publication, with due recognition of subsequent changes, and are to be thought of as the official American position on nomenclature matters.

The names and symbols in Table 4–1 are the official designations for the elements. It would be surprising to find substantial differences between the list in Table 4-1 and those lists of elements given in textbooks printed after 1965. These names and symbols are well established. It is assumed, at the least, that you can quote from memory the trivial names and symbols for the elements with atomic numbers 1 through 40.

4-T-1

You should also commit to memory the trivial names, symbols, and Latin names for copper, gold, iron, lead, silver, and tin. Each symbol was derived from the Latin names, as you can verify.

The Latin name is used when forming names derived from these elements: for example, ferrate, cuprate, and stannate are used rather than ironate, copperate, and tinate, respectively.

The observant student will also notice that there are five other elements with symbols that do not correlate with their trivial names: antimony (Sb), mercury (Hg), potassium (K), sodium (Na), and tungsten (W). Each of these symbols was derived from an alternative name that has fallen into disfavor, except for wolfram, which is recommended by the IUPAC. Tungsten is preferred in English-speaking countries.

Table 4-1. List of Elements and Their Symbols

Name	Symbol	Name	Symbol	Name	Symbol
Actinium	Ac	Hafnium	Hf	Protactinium	Pa
Aluminum	Al	Helium	He	Radium	Ra
Americium	Am	Holmium	Ho	Radon	Rn
Antimony	Sb	Hydrogen	H	Rhenium	Re
Argon	Ar	Indium	In	Rhodium	Rh
Arsenic	As	Iodine	I	Rubidium	Rb
Astatine	At	Iridium	Ir	Ruthenium	Ru
Barium	Ba	Iron (ferrum)	Fe	Samarium	Sm
Berkelium	Bk	Krypton	Kr	Scandium	Sc
Beryllium	Be	Lanthanum	La	Selenium	Se
Bismuth	Bi	Lead (plumbum)	Pb	Silicon	Si
Boron	B	Lithium	Li	Silver (argentum)	Ag
Bromine	Br	Lutetium	Lu	Sodium	Na
Cadmium	Cd	Magnesium	Mg	Strontium	Sr
Calcium	Ca	Manganese	Mn	Sulfur	S
Californium	Cf	Mendelevium	Md	Tantalum	Ta
Carbon	C	Mercury	Hg	Technetium	Tc
Cerium	Ce	Molybdenum	Mo	Tellurium	Te
Cesium	Cs	Neodymium	Nd	Terbium	Tb
Chlorine	Cl	Neon	Ne	Thallium	Tl
Chromium	Cr	Neptunium	Np	Thorium	Th
Cobalt	Co	Nickel	Ni	Thulium	Tm
Copper (cuprum)	Cu	Niobium	Nb	Tin (stannum)	Sn
Curium	Cm	Nitrogen	N	Titanium	Ti
Dysprosium	Dy	Nobelium	No	Tungsten	
Einsteinium	Es	Osmium	Os	(wolfram)	W
Erbium	Er	Oxygen	O	Uranium	U
Europium	Eu	Palladium	Pd	Vanadium	V
Fermium	Fm	Phosphorus	P	Xenon	Xe
Fluorine	F	Platinum	Pt	Ytterbium	Yb
Francium	Fr	Plutonium	Pu	Yttrium	Y
Gadolinium	Gd	Polonium	Po	Zinc	Zn
Gallium	Ga	Potassium	K	Zirconium	Zr
Germanium	Ge	Praseodymium	Pr		
Gold (aurum)	Au	Promethium	Pm		

Q1. What is the Latin name for copper?

A1. cuprum

Q2. What is the symbol for copper: Cu, Co, or Cr?

A2. Cu

Q3. Ag is the symbol for _____,
which has the Latin name argentum.

A3. silver

Q4. The trivial name for the element
represented by Sn is _____.

A4. tin

Q5. Hexacyanoferrate(III) ion contains
which metal?

A5. iron

Q6. $Pb(NO_3)_2$ contains which metal?

A6. lead

Q7. Potassium tetrahydroxoaurate contains which two metals?

A7. potassium and gold

Q8. What is the symbol for sodium: S, So,
Na, Sd, or Nt?

A8. Na

Q9. Sb represents _____.

A9. antimony

Q10. Mercury has the symbol Me.
true/false

A10. false; the correct
symbol is Hg

The mass number, atomic number, number of atoms, and ionic charge of an element is to be indicated by means of four indices placed around the symbol. The positions are to be occupied thus:

$$\begin{matrix} \text{mass number} & & \text{ionic charge} \\ \text{atomic number} & \text{A} & \text{number of atoms} \end{matrix}$$

The ionic charge should be indicated by A^{n+} rather than A^{+n}.*

Example: $^{200}_{80}Hg_2^{2+}$ represents the doubly charged ion containing two mercury atoms, each of which has the mass number of 200.

In common practice, the only time the mass number and atomic number are written explicitly is in work where knowledge of the exact isotope is critical, such as in nuclear chemistry and radioactive isotope reactions. When no mass number is written, you assume that the mass number appearing in an atomic weight table is meant. This, as you may know, is a weighted average based upon the relative abundance of the naturally occurring isotopes of the element.

Example: O_3^{2+} represents the doubly charged ozone ion, consisting of three oxygen atoms each of mass number 16.

It should also be noted that when the charge is 1+ or 1−, and when a particle consists of a single atom, the numeral 1 is not written.

Q1. Where should the ionic charge be placed on the atomic symbol?

A1. upper right index

Q2. Should the charge be written $n+$ or $+n$?

*"Nomenclature of Inorganic Chemistry," *J. Am. Chem. Soc.,* **82,** 5526 (1960).

A2. $n+$

Q3. The atomic number is properly placed in the _____ corner.

A3. lower left

Q4. What is incorrect about this symbol: $^{7}_{14}N^{3+}$?

A4. atomic and mass numbers are in wrong positions

Q5. Does the symbol $^{18}_{9}F^{-1}$ follow the IUPAC recommendations?

A5. no; the charge is written incorrectly

Q6. What number should appear in the lower right index?

A6. number of atoms in the species

Q7. The symbol Na^+ is acceptable for a singly charged sodium ion. What mass number is implied?

A7. 23 (see the atomic weight table)

Q8. What does the formula $^{18}F^{19}F$ represent?

A8. a diatomic molecule consisting of one atom of fluorine mass number 18 and one atom of fluorine mass number 19; the charge is zero

Q9. Elemental sulfur has the symbol S_8.
What does this mean?

A9. elemental sulfur is
found in a molecule
consisting of eight
sulfur atoms; the
mass number of each
atom is assumed to
be 32

SUMMARY

The IUPAC, supported by the American Chemical Society and the National Research Council, has made very specific recommendations regarding the names and symbols of the elements and the use of the four indices surrounding an atomic symbol.

Six elements, in addition to their trivial names, have Latin names that are used in naming compounds of these elements, and five other elements have symbols derived from names that are not commonly used today.

The four indices surrounding an atomic symbol are supposed to be used as follows:

$$\begin{matrix} \text{mass number} \\ \text{atomic number} \end{matrix} A \begin{matrix} \text{ionic charge} \\ \text{number of atoms} \end{matrix}$$

In practice, it is common to omit the mass number and atomic number except in situations where it is essential to know the exact isotope, such as in nuclear chemistry.

The knowledgeable student can translate the atomic symbol, in its various forms, into words. Further, he or she can recognize and use the Latin or alternative names and symbols for the appropriate elements as well as the trivial names and correct symbols for most of the common elements in the periodic table.

1. (3 pt) Give the correct trivial name and Latin name, where appropriate, for each of the following: Zn, Mn, Cl, Sn, Au, W.

2. (3 pt) Write the correct symbols for silver, lead, sodium, potassium, selenium, and chromium.

3. (2 pt) Is the following representation for lithium correct: $^3_7L^{+1}$? If not, point out its deficiencies.

4. (2 pt) What is the meaning of the formula $^2H^{37}Cl$?

5

FORMULAS

OBJECTIVES

The student should be able to

1. Differentiate among the empirical, the molecular, and the structural formula for a compound by using an example and a brief statement.
2. Predict the most likely empirical formula for a simple binary compound or a compound using common complex ions, given the oxidation states of its components.
3. Use the rule and its major exceptions for writing formulas for ternary, and larger, compounds.

1. (2 pt) The following information on a compound was obtained by laboratory analysis: atom ratio, one carbon to three hydrogens; molecular weight, 30. What is the (a) empirical formula and (b) molecular formula for this compound?

2. (1 pt) The following are all representations of cyclopropane. Which one is the structural formula?

(a) C_3H_6

(b) $CH_2CH_2CH_2$

(c) $CH_2\!-\!CH_2$
        $\diagdown\diagup$
        CH_2

3. (1 pt) When writing an empirical or molecular formula, should the more electropositive or the more electronegative constituent be written first?

4. (1 pt) Given that the oxide ion has an oxidation number of 2− and that the gallium ion is Ga^{3+}, predict the formula of gallium oxide.

5. (1 pt) Phosphate is an anion represented by PO_4^{3-}, and calcium, being a member of the alkaline earth metals, has an expected oxidation number of 2+. Predict the empirical formula of the compound formed by these two species.

6. (1 pt) When the two nonmetals nitrogen and oxygen combine, they form several compounds. Which element should be written first in the formulas for these compounds?

7. (2 pt) What is the general rule to be followed when writing the formula of a ternary compound? There are two important classes of compounds that are exceptions to this rule. Give them.

8. (1 pt) Which element should be written first in a compound between hydrogen and boron?

INTRODUCTION

A formula is a simple means of representing a compound. It gives us information about the composition and, in some cases, the structure of the compound in a compact and efficient fashion. It is usually based upon a laboratory analysis, although predictions for formulas based upon commonly known oxidation states and valence can be accurate. You are likely to encounter three types of formulas in your study of chemistry. They are (1) simplest or empirical formulas, (2) molecular formulas, and (3) structural formulas.

Structural formulas indicate the sequence and spatial arrangement of the atoms in a molecule. These result from specific structural studies using various instruments and methods of physical chemistry. Examples of structural formulas are

$$
\begin{array}{ccc}
& \text{H} & & & \text{O} \\
& | & & & | \\
\text{H}-&\text{C}-\text{H} & \text{and} & \text{H}-\text{O}-&\text{S}-\text{O}-\text{H} \\
& | & & & | \\
& \text{H} & & & \text{O}
\end{array}
$$

Frequently, these arrangements are expressed in three-dimensional models.

An **empirical formula** gives the lowest possible whole number of each kind of atom that is consistent with the composition of the substance. On the other hand, a **molecular formula** expresses the actual, rather than lowest, number of each kind of atom in the molecule. These are determined from the molecular weight of the compound. To illustrate, chemical analysis of a certain sulfur–chlorine compound gave a ratio of one sulfur atom to one chlorine atom. The molecular weight of this compound was found to be 135. The empirical formula is SCl and the molecular formula is S_2Cl_2. The first formula gives the lowest reasonable number of atoms consistent with the composition; the second formula maintains the correct ratio and also corresponds to the experimental molecular weight:

$$
\begin{array}{rl}
2 \times 32 & = 64\ S_2 \\
2 \times 35.5 & = \underline{71\ Cl_2} \\
& 135\ S_2Cl_2
\end{array}
$$

(Note that S_3Cl_3 and S_4Cl_4 also show the 1:1 ratio for sulfur and chlorine, but they fail to give the correct molecular weight and are thereby ruled out.) We would not choose fractional numbers because only whole atoms combine to form molecules or ions.

Finally, two more points in this brief review: (1) when the molecular weight varies with the conditions (e.g., temperature), the simplest formula is often chosen to represent the substance. For example, S may be used instead of S_8 for sulfur; (2) in most cases where the ions or atoms of a solid are bound together in one huge aggregate or molecule, we use the empirical formula to represent the solid because the molecular formula is unwieldy and cannot be easily used to balance equations.

Examples: NaCl instead of $Na_n Cl_n$ for sodium chloride, an ionic solid; Cu instead of Cu_n for metallic copper, and C instead of C_n for graphite (one solid form of carbon), a covalent network solid.

5-T-1

Even though many modern laboratory courses ask the student to determine the simplest or empirical formula of a simple compound, it is obvious that you will not go through the experimental routine for 99.99 percent of the compounds you will encounter in your study of chemistry; therefore, you need to develop skill in predicting formulas for various combinations of elements. The rules you need are few and simple. They are:

1. The electropositive constituent (cation), as a general rule, is written first in empirical and molecular formulas.
2. The net charge on a multinuclear species is the sum of the oxidation numbers of its constituent atoms. Compounds always have net charges of zero.
3. Knowledge of the oxidation numbers of the species involved in the substance.

The process of predicting and writing the formula consists quite simply of putting the positive and negative constituents together in such a way as to obey statements 1 and 2 above. To illustrate, suppose you are told that aluminum has an oxidation state of 3+ and that chlorine is 1- and are asked to predict the formula for the simplest compound formed by these two elements. You proceed by applying the principle that the net charge should be zero for a compound (Rule 2 above). The 3+ brought to the compound by Al should be balanced exactly by the chloride ions; thus we require three Cl^-. The formula gives the number of each atom in the lower right position (subscript) with no number understood to mean 1, and since aluminum is the more electropositive (Rule 1), its symbol is written first; thus the correct formula in this case is $AlCl_3$.

In a slightly more difficult example, let us predict the formula for the compound, aluminum sulfate, given that the net charge of the sulfate anion is 2-. If these ions are combined in a 1:1 manner, we will get $AlSO_4^+$, which is not a compound formula because its net charge is not zero. We must find the smallest whole number for each constituent in a combination that will yield a net charge of zero. This is usually done in a systematic trial-and-error procedure. Suppose that we choose two Al^{3+}; then we must balance 6+, which can be done with three sulfate ions, which contribute a total of 6-. The simplest formula would be $Al_2(SO_4)_3$. Once again, aluminum is written first because it is the cation. Note that the 3 appears as a subscript outside the parentheses surrounding the SO_4 unit. $Al_2S_3O_{12}$ would not represent aluminum sulfate as clearly, even though it has the correct molecular weight, because it does not show any special relationship between the sulfur and oxygen atoms when, in fact, there is one. The parentheses are used to show that units of one sulfur bound to four oxygen atoms are in turn, bound to an aluminum ion, rather than some other combination of sulfur and oxygen atoms.

Returning to the procedure, if our try with two Al^{3+} had not produced a balanced total charge, we would then try three Al^{3+}, and so forth. Remember that we are looking for the smallest whole numbers, so we always start with one and increase by one until the proper combination is found.

Q1. The simplest formula for a molecule is also known as the _____ formula.

A1. empirical

Q2. The molecular formula is a simple multiple of the empirical formula. true/false

A2. true

Q3. The simplest formula for a certain substance is A_2B. This means there are _____ atoms of A per atom of B in this substance.

A3. two

Q4. In Q3, how many atoms of B are there per atom of A?

A4. one-half

Q5. The net charge on a molecule is zero. true/false

A5. true

Q6. A certain molecule is represented by the formula $A_2B_3^{2+}$. Is this correct?

A6. no; molecules do not have net charges

Q7. Suppose that you encountered the formula P_4^0. Is this a molecule?

A7. yes; the superscript 0 together with the subscript 4 are the clues

Q8. Predict the formula for the compound formed between M^{2+} and D^{3-}.

A8. M_3D_2

Q9. What formula would you predict for M^{2+} and X, where $X = Cl^-, Br^-, I^-$?

A9. MX_2

Q10. The phosphate ion is PO_4^{3-}. What is the charge on this ion?

A10. 3−

Q11. Predict the formula of sodium phosphate.

A11. Na_3PO_4 (remember, group 1A has usual oxidation numbers of 1+)

Q12. Write the formula for calcium phosphate.

A12. $Ca_3(PO_4)_2$ (don't forget the parentheses)

Q13. If gold has two likely oxidation states, 1+ and 3+, what would you predict for the formulas of the oxides of gold?

A13. Au_2O, Au_2O_3

Q14. Chlorine is found with an oxidation number of 1+ in the hypochlorous ion, ClO^-. Is this the correct charge for this ion, assuming normal behavior for the oxygen?

A14. yes: O = 2−
$\underline{\qquad Cl = 1+}$
net = 1−

Q15. Predict the formula for a molecule formed between ClO^- and an alkaline earth metal M.

A15. $M(ClO)_2$ (parentheses are essential; $MClO_2$ represents a different substance)

Q16. The more electronegative constituent appears first in the molecular formula. true/false

A16. false

Q17. M is a metal, O is a nonmetal. Which should be written first in a formula representing a compound of these two?

A17. M; metals usually are more electropositive than nonmetals

5-T-2

There are *two* important *exceptions* to the formula rule that states that the more electropositive element is always written first. They are found in formulas of

1. Binary compounds between *nonmetals.*
2. Compounds containing three or more elements.

The rule for binary compounds between nonmetals is that the constituent that appears first in the following list is also written first in the formula: B, Si, C, Sb, As, P, N, H, Te, Se, S, I, Br, Cl, O, F. The list roughly parallels an order of increasing electronegativity. Unfortunately, there are some elements out of order, the most notable of which is hydrogen.

Q1. The least electronegative element is always written first in a formula. true/false

A1. false

Q2. Two exceptions exist to the rule of writing the most electropositive element first. One involves ternary or

larger compounds and the other treats
_____ compounds be-
tween _____ .

A2. binary, nonmetals

Q3. Which element should be written first
in a formula representing a compound
between
(a) H and F
(b) B and N
(c) Si and C
(d) Br and Cl
(e) C and H

A3. (a) H
(b) B
(c) Si
(d) Br
(e) C

Q4. Ammonia contains three hydrogen
atoms for each nitrogen atom. Which
is the proper formula for this compound, H_3N or NH_3?

A4. NH_3

Q5. An analysis of an interhalogen *compound* between iodine and chlorine
gave an atomic ratio of 1:1. What is its
empirical formula?

A5. ICl

Q6. A phosphorous compound analogous
to ammonia has been made. Predict
its formula.

A6. PH_3 (it has a trivial
name of phosphine)

Q7. Suppose that you made a compound consisting of arsenic and sulfur. Which symbol should appear first in the formula?

A7. As

Q8. Is the formula Cl_2O written correctly with respect to order of atoms?

A8. yes

5-T-3

Compounds and ions that contain three or more elements follow the rule that the sequence of atoms in the formula be the same as the order in which the atoms are actually bound in the molecule. To illustrate, laboratory work has demonstrated that the sequence of atoms in the cyanate ion is OCN^-. If you carelessly or unknowingly wrote ONC^-, you would, in fact, be representing an altogether different species, the fulminate ion. As you can see, it can be very important to know the proper atomic sequence. These must simply be memorized. Fortunately, there are not many of these cases.

Acids containing oxygen such as H_2SO_4 and $HClO_4$ do not follow the sequence rule either. The IUPAC Commission allows this exception because it reflects a long-standing practice that they do not wish to disturb. You can usually recognize these formulas because the acidic hydrogen is written first. A separate chapter will deal more fully with these compounds.

Another important exception to the rule are the formulas for many organic compounds. It is customary in organic chemistry formulas to write C, H, and then the other elements, regardless of the actual bonding sequence. Structural formulas are then used to show the actual sequence or arrangement of the atoms.

Q1. In a formula for a compound containing three or more elements, recognizing the exceptions, the atom order should reflect the actual bonding sequence. true/false

A1. true

Q2. Structural studies show that the atoms in the thiocyanate ion are bound in the order SCN^-. Would NSC^- also represent this ion?

A2. no

Q3. The sequence of atoms in HNO_3 is likely to represent the actual bonding sequence. true/false

A3. false

Q4. The structure of perchloric acid is

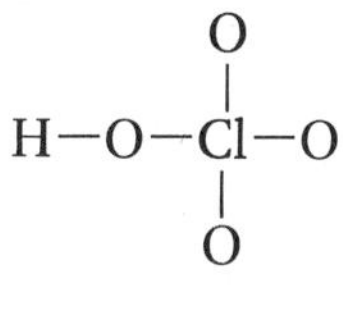

$$H-O-\overset{\overset{\displaystyle O}{|}}{\underset{\underset{\displaystyle O}{|}}{Cl}}-O$$

Would the formula $HOClO_3$ essentially follow the rule of this section?

A4. yes

Q5. Dimethyl ether can be represented by three different-appearing formulas. Which comes closest to the rule of this section?
(a) $C_2 H_6 O$
(b) $H_3 COCH_3$
(c)

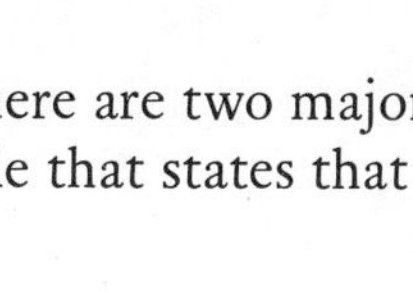

$$H-\overset{\overset{\displaystyle H}{|}}{\underset{\underset{\displaystyle H}{|}}{C}}-O-\overset{\overset{\displaystyle H}{|}}{\underset{\underset{\displaystyle H}{|}}{C}}-H$$

A5. (b)

Q6. There are two major exceptions to the rule that states that the bonding sequence should be shown in the formula. They are _____ and _____ .

A6. oxyacids, organic
      compounds

SUMMARY

There are three types of formulas encountered in chemistry: empirical, molecular, and structural. All are based on information obtained by experiments. We can frequently predict formulas if we know (1) the oxidation numbers of the participating species, (2) the net charge on the required species, and (3) that the most electropositive constituent is written first (with certain exceptions).

1. (1 pt)   A certain compound containing aluminum and sulfur was subjected to laboratory analysis and the following data were collected: molecular weight, 150; atomic ratio, Al/S = 1:1½. What is the molecular formula for this compound?

2. (2 pt)   Using an example of your choice, differentiate among the empirical, molecular, and structural formulas of a compound.

3. (2 pt)   The following nonmetals are to be arranged in the IUPAC order used in determining which is written first in a formula of a compound between any two of them: B, O, H, F, N, I.

4. (1 pt)   Predict the formula for calcium oxide.

5. (1 pt)   An interhalogen compound between bromine and iodine has been made that has a simple 1:1 atomic ratio. Predict the empirical formula for this compound.

6. (1 pt)   Nitric acid is represented by the formula HNO_3. Briefly tell how this deviates from the formula-writing rule for ternary compounds.

7. (1 pt)   What would be the predicted empirical formula of the compound formed between gallium (Ga^{3+}) and the silicate anion, SiO_4^{4-}?

8. (1 pt)   NaCNO and NaNCO have the same molecular weight. Strictly speaking, are these two representations equivalent? Explain.

6

BINARY COMPOUNDS

OBJECTIVES

The student should be able to

1. Give a definition of a binary compound that includes at least two examples.
2. Compose the recommended IUPAC name for a compound formed between any two common elements and, if given a periodic table, write the name for any reasonable binary compound.
3. Recognize and use alternative methods for naming binary compounds when more than one compound is formed between two elements.
4. Write the formula for a binary compound given its name in any of the commonly accepted forms of inorganic nomenclature.
5. State from memory the acceptable trivial name for each of the following hydrides: H_2O, NH_3, N_2H_4, B_2H_6, SiH_4, PH_3, AsH_3, SbH_3.

1. (2 pt)   Which of the following are binary compounds?
   (a) $Ca(OH)_2$ (b) N_2O
   (c) $NaBr$ (d) H_2SO_4

2. (9 pt)   Write a suitable IUPAC name for
   (a) Ag_2S (b) TiI_3
   (c) B_2H_6 (d) ZnO
   (e) Al_2Se_3 (f) PBr_5
   (g) NO_2 (h) ICl
   (i) NH_3

3. (2 pt)   Differentiate between the compounds Cu_2O and CuO by using the Stock system of naming.

4. (2 pt)   Using prefixes, distinguish by name between VF_2 and VF_3.

5. (2 pt)   Using endings, give distinguishing names for $FeCl_2$ and $FeCl_3$.

6. (8 pt)   Write the formula for the following compounds:
   (a) sodium nitride
   (b) calcium carbide
   (c) cobaltic oxide (cobalt may be either 2+ or 3+)
   (d) manganese(IV) sulfide
   (e) dinitrogen tetraoxide
   (f) barium oxide
   (g) silane
   (h) ammonia

INTRODUCTION

We have arrived at the point where the main purpose of this book comes into focus. The material in the preceding chapters was preliminary in the sense that it was needed to develop terms, expectations, and conventions that, in many cases, are essential to an understanding of the rules of inorganic nomenclature.

The nomenclature of five types of substances will be discussed. We begin with the simplest of these.

6-T-1

Compounds formed by the combination of two elements are called **binary compounds.** Sometimes there may be more than one atom of each of the two substances, such as in A_2B_3. These compounds may be ionic, in which case their names and formulas correspond to the simplest stoichiometric composition of the substances, or they may be molecular.

> **Q1.** A binary compound contains just _____ elements.

A1. two

> **Q2.** Which of the following are binary compounds? NaBr, H_2S, CCl_4, MgSe, BF_3.

A2. all of them

6-T-2

Preferred IUPAC names are formed for binary compounds by indicating the elements and their proportions according to the following rules:*

1. The name of the electropositive element, or the element that is treated like one, is written first without any modification.

*"Nomenclature of Inorganic Chemistry," *J. Am. Chem. Soc.,* **82,** 5527 (1960).

2. The name of the electronegative constituent is written second and is modified to end in -ide. Usually, this modification is accomplished by stripping the name of the element back to its next-to-last consonant and then adding -ide: for example, sulfur, sulf, sulfide; chlorine, chlor, chloride. The following exceptions need to be memorized: oxygen, oxide; nitrogen, nitride; hydrogen, hydride; and phosphorus, phosphide.

For binary compounds between nonmetals, the elements standing later in the list given in Section 5-T-2 are modified to end in -ide.

Certain polyatomic groups are also given the ending -ide. (See Section 7-T-4 for more on this.)

Q1. When writing the name of a binary compound, the proper order gives the more _____ element first.

A1. electropositive

Q2. The name of the more electropositive element is modified. true/false

A2. false

Q3. When modifying the name of the electronegative constituent in a binary compound, you simply drop the last syllable and add -ide. true/false

A3. false

Q4. The proper way to modify the name of the electronegative constituent in a binary compound is to strip it back to the _____ consonant and then add _____.

A4. next-to-last, -ide

Q5. The suffix or ending is used without a hyphen. true/false

A5. true

Q6. Two elements that are exceptions to the rule stated in Q4 are _____ and _____.

A6. oxygen, oxide
nitrogen, nitride
hydrogen, hydride
phosphorus, phosphide

Q7. From a list of nonmetals we find the following order: B, Si, C, H. Suppose that a 1:1 compound is formed between boron and silicon. When naming:
(a) Which element is written first?
(b) Should its name be modified? If so, how?

A7. (a) boron
(b) no

Q8. Referring again to the list in Q7, suppose that the 1:1 compound was formed between silicon and carbon. When naming:
(a) Which element is written second?
(b) Should it be modified? If so, how?

A8. (a) carbon
(b) yes; carbon to carbide

Q9. Give the IUPAC preferred names for the following: NaCl, CaS, MgTe, SiC.

A9. NaCl, sodium chloride
CaS, calcium sulfide
MgTe, magnesium telluride
SiC, silicon carbide

Q10. Criticize the following names and then give the correct name:
(a) FeC, ferrum carbide
(b) AlN, aluminum nitrogide
(c) NaBr, sodide bromine

A10. (a) should not use the Latin name for iron; iron carbide
(b) modified nitrogen incorrectly; aluminum nitride
(c) modified the ending of the wrong element; sodium bromide

Q11. Write the formula for the name
(a) lithium iodide
(b) calcium oxide
(c) radium selenide
(d) zinc sulfide

A11. (a) LiI
(b) CaO
(c) RaSe
(d) ZnS

6-T-3

Proportions of the various parts of a compound are given by Greek numerical prefixes. These follow in numerical order: mono, di, tri, tetra, penta, hexa, hepta, octa, ennea, deca. In practice, the Latin prefix for nine, nona, is used instead of ennea, and the prefix mono is usually omitted. The prefix precedes the name of the element or group without a hyphen.

Examples illustrating the use of prefixes are N_2O, dinitrogen oxide; N_2O_4, dinitrogen tetraoxide; CCl_4, carbon tetrachloride; S_2Cl_2, disulfur dichloride; and PBr_5, phosphorus pentabromide.

Q1. The proportions of the constituents in a compound are expressed using Greek or Latin _____.

A1. prefixes

Q2. The Greek prefix corresponding to each of the following is
(a) three   (b) five
(c) four   (d) six

A2. (a) tri
(b) penta
(c) tetra
(d) hexa

Q3. The prefix usually used for nine is

_____ .

A3. nona

Q4. The prefix is used with/without a hyphen.

A4. without

Q5. A compound has the formula NO_2. Criticize the name nitrogen oxide (di).

A5. the prefix is not used properly: it should precede the name of the element with no parentheses; nitrogen dioxide is correct

6-T-4

As mentioned earlier, the prefix mono is rarely used. Other prefixes can also be omitted if there is no chance of misunderstanding the proportions of constituents in the compound. To illustrate, the name "calcium chloride" is commonly used for $CaCl_2$. Strictly speaking, the name should be calcium dichloride. Why is this deviation from the nomenclature rules acceptable?

The answer is that people with a good knowledge of the periodic chart will know the most likely oxidation states of calcium and chlorine by virtue of their respective positions in the chart and thus infer the correct formula. There are two considerations involved: (a) calcium and chlorine are relatively far apart in electronegativity and thus will form an ionic compound; and (b) because calcium belongs to 2A and chlorine is in 7A, we should recall that their most common oxidation states will be 2+ and 1–, respectively. One thus infers that the formula for calcium chloride is $CaCl_2$, even though the proportions are not specifically given in the name. This approach is used for many compounds, and it frequently presents a puzzling situation to beginning students who try to strictly follow the rules. Some hints to aid you with these might be helpful. First, learn the principal oxidation state associated with each major family, the members of each family, and their relative positions on the periodic chart. Second, when you see a name that lacks prefixes, do not automatically assume that only mono has been omitted. Make a quick mental check, or look at a periodic chart, and see what families the elements fall into. If the oxidation states differ numerically, you will have to use the charge balance principle to work out the formula.

The following examples illustrate cases where the prefixes are often omitted: calcium oxide, both elements have same oxidation state except for the sign, thus CaO; sodium sulfide, sodium found in group 1A, sulfur from group 6A; their respective oxidation states are 1+ and 2–, thus Na_2S; magnesium nitride, magnesium is a member of group 2A and nitrogen belongs to 5A; their respective oxidation states are expected to be 2+ and 3–, thus Mg_3N_2.

Q1. Which numerical prefix is commonly dropped?

A1. mono

Q2. You can drop other numerical prefixes whenever the name is without ambiguities. true/false

A2. true

Q3. Lithium oxide is an acceptable name for Li_2O. true/false

A3. true; this is a compound in which the oxidation states are predictable

Q4. The oxidation state of Be in BeO is 2+; therefore, the only name for BeO should be diberyllium dioxide. true/false

A4. false; the 2+ state is exactly balanced by the 2− for oxygen; thus the proportion is 1:1 and an acceptable name is beryllium oxide

Q5. Write the principal oxidation state for the following families:
(a) 7A
(b) 5A
(c) 3A

A5. (a) 1−
(b) 3−
(c) 3+

Q6. Potassium falls into 1A and sulfur in 6A. What is the expected formula for potassium sulfide?

A6. K_2S

Q7. Write an acceptable name for
(a) $CaBr_2$ (b) Ba_3N_2

A7. (a) calcium bromide or calcium dibromide
(b) barium nitride or tribarium dinitride

Q8. Write an acceptable name for
(a) AlN (b) Li_2S

A8. (a) aluminum nitride
(b) lithium sulfide or dilithium sulfide

Q9. Write an acceptable name for
(a) K_2Se (b) RbI

A9. (a) *potassium selenide or dipotassium selenide*
(b) *rubidium iodide*

Q10. Write the simplest formula for
(a) sodium chloride
(b) potassium oxide

A10. (a) NaCl
(b) K_2O

Q11. Write the simplest formula for
(a) calcium fluoride
(b) magnesium oxide
(c) aluminum chloride

A11. (a) CaF_2
(b) MgO
(c) $AlCl_3$

Q12. An acceptable name for Al_2O_3 is aluminum oxide. The basic reason we do not need to show prefixes in this name is because chemists should know the primary oxidation states for Al and O. true/false

A12. true

6-T-5

The IUPAC states that prefixes are acceptable in the name of any compound *but* they are especially suitable when naming compounds between nonmetals. You should take this to mean that the prefixes, excepting mono, should almost always be used when naming compounds between nonmetals. This position is taken because it is very likely that more than one compound will be formed between two given nonmetals, thus a name without prefixes could easily represent two or more compounds.

Examine the situation between nitrogen and oxygen. These two nonmetals can form, depending on the conditions, any one of the following binary compounds: N_2O, NO, N_2O_3, NO_2, N_2O_4, N_2O_5, NO_3, and N_2O_6. It is immediately obvious that the name nitrogen oxide, taken in the sense of the last section, does not distinguish between these. In the strict sense, it only represents NO.

Q1. The preferred IUPAC name for binary compounds between nonmetals includes _____ whenever applicable.

A1. prefixes

Q2. The nonmetals are found in the _____ right corner of the periodic chart.

A2. upper

Q3. Write the preferred names for
(a) N_2O_5 (b) NO_2

A3. (a) dinitrogen pentaoxide
(b) nitrogen dioxide

Q4. Write the preferred names for
(a) IBr (b) NF_3

A4. (a) iodine monobromide
(b) nitrogen trifluoride

Q5. Write the preferred names for
(a) P_4S_3 (b) Si_2Cl_6

A5. (a) tetraphosphorus trisulfide
(b) disilicon hexachloride

Q6. Write suitable names for

(a) OF_2 (b) SO_3

A6. (a) oxygen difluoride
(b) sulfur trioxide

Q7. Write the formula for

(a) diiodine pentaoxide
(b) hydrogen chloride
(c) dihydrogen oxide
(d) tellurium tetrachloride

A7. (a) I_2O_5
(b) HCl
(c) H_2O
(d) $TeCl_4$

6-T-6

The proportions of the constituents of a compound may also be expressed indirectly by the **Stock system**. The essential feature of the Stock system is that it uses a Roman numeral to represent the oxidation state or valence of an element. This numeral is placed in parentheses immediately following the name. The arabic 0 is used for zero. You do not use prefixes. In practice, the 1+ or 1– oxidation state is rarely indicated using a Roman numeral.

Examples using the Stock system are MnO_2, manganese(IV) oxide; CuF_2, copper(II) fluoride; and Al_2S_3, aluminum(III) sulfide. Note that the rules for modifying the names of the elements remain the same. The only new features are (1) dropping the use of prefixes, and (2) insertion of the parenthetical Roman numeral between the constituents.

The IUPAC prefers that when using the Stock system, the Latin name for an element be used when it is given, such as ferrum(II) chloride for $FeCl_2$ instead of iron(II) chloride. The use of the Latin name has not caught on, especially in the United States, and you are unlikely to encounter this usage.

There are two other points to be emphasized about the Stock system. They are: (1) it can be applied to both cations and anions, and

(2) the IUPAC prefers that it not be applied to compounds between nonmetals. As an example of point 2, unpreferred usage of the Stock system would be nitrogen(V) oxide for N_2O_5, rather than dinitrogen pentaoxide.

Q1. The Stock system is another way of expressing the proportions of constituents in a compound. true/false

A1. true

Q2. The Stock system uses ——————— numerals to express the oxidation state of an element.

A2. Roman

Q3. The Roman numeral is placed in parentheses immediately following the name of the compound/element.

A3. element

Q4. The Stock system should be used with compounds made of nonmetals. true/false

A4. false

Q5. The preferred method of indicating the proportions of elements in compounds composed only of nonmetals is by the use of ———————.

A5. prefixes

Q6. Use the Stock system to write a suitable name for
(a) $TiCl_3$ (b) ZnSe

A6. (a) titanium(III) chloride
(b) zinc(II) selenide

Q7. Write a Stock system name for
(a) Cr_2O_3 (b) Al_4C_3

A7. (a) chromium(III) oxide
(b) aluminum(III) carbide

Q8. Write suitable names for
(a) PCl_5 (b) Ga_2O_3

A8. (a) phosphorus pentachloride
(Stock system not used
because both elements
are nonmetals)
(b) gallium(III) oxide

Q9. Write suitable Stock names for
(a) Fe_2O_3 (b) $CuBr_2$

A9. (a) ferrum(III) oxide or
iron(III) oxide
(b) cuprum(II) bromide or
copper(II) bromide

6-T-7

A third system for indirectly indicating the proportions of the elements in a compound is also in use. It is not recommended by the IUPAC and should not be encouraged. It is mentioned here only because you will probably encounter it and you should be able to interpret it.

The system adds the suffixes -ous and -ic, depending upon the oxidation state, to the root of the name of the more electropositive element. The Latin name is often used. The -ous ending is used for the lower of two oxidation states and the -ic is for the higher oxidation state. Thus, in the case of CuCl and $CuCl_2$, the names, using this system, would be cuprous chloride and cupric chloride, respectively. The Roman numerals of the Stock system and Greek prefixes are not used. The ending of the name for the electronegative constituent is formed as before.

The system requires that you know the two oxidation states, and whenever three or more oxidation states are possible, the scheme runs into serious difficulty unless other prefixes and endings are introduced.

Vanadium, for example, can be found in 2+, 3+, 4+, and 5+ oxidation states. It should be obvious that vanadous and vanadic will not clearly indicate which two states are involved, although there are some who assign the endings to the first two in the series. The practice is inconsistent and for this reason is being discouraged. It is most effective only when the electropositive element has just two oxidation states.

Q1. The ending -ic, when added to the electropositive element's name, indicates the higher of two oxidation states. true/false

A1. true

Q2. If tin has oxidation states of 2+ and 4+, the correct formula for stannous oxide is _____ .

A2. SnO

Q3. Given that lead exists in either 2+ or 4+ oxidation states, write the name for PbS_2 using the -ous or -ic system.

A3. plumbic sulfide

Q3. Auric chloride is represented by the formula $AuCl_3$. What would be likely oxidation states for gold in aurous chloride?

A4. either 2+ or 1+; experimentally there is little, if any, evidence for Au^{2+}

Q5. Iron shows oxidation states of 2+ and 3+. Write formulas and -ous/-ic names for its compounds with bromine.

A5. $FeBr_2$, ferrous bromide
$FeBr_3$, ferric bromide

Q6. Write -ous/-ic names for Hg_2S and HgS (assuming no other oxidation states for mercury).

A6. mercurous sulfide, Hg_2S
mercuric sulfide, HgS

6-T-8

In spite of all attempts to build a systematic nomenclature, there are some trivial names for compounds that persist and are acceptable. You probably already know at least two of these and the others need to be memorized. They are:

H_2O	water	SiH_4	silane
NH_3	ammonia	PH_3	phosphine
N_2H_4	hydrazine	AsH_3	arsine
B_2H_6	diborane	SbH_3	stibine

The trivial name is suitable, although you would not be wrong to give a systematic name, such as silicon tetrahydride, if you so choose. Be sure you can also write the formula if given the trivial name.

Q1. Are trivial names as opposed to systematic names acceptable in the IUPAC scheme?

A1. yes, but only for a very limited number of compounds

Q2. Give the trivial names for
(a) NH_3 (b) PH_3
(c) AsH_3 (d) SbH_3

A2. (a) ammonia
(b) phosphine
(c) arsine
(d) stibine

Q3. Carefully examine the formulas for the compounds in Q2. What do they hold in common?

A3. they all have the same
general formula AH_3; the
elements represented by
A all belong to family 5A

Q4. What connection can you see between
the names and the generalizations you
found in Q3?

A4. excepting ammonia, the
names all have the same
ending, -ine

Q5. Write the trivial name for
(a) B_2H_6 (b) N_2H_4

A5. (a) diborane
(b) hydrazine

Q6. Write the formula for
(a) silane   (b) ammonia

A6. (a) SiH_4
(b) NH_3

Q7. If diarsine is represented by As_2H_4,
predict the formula for diphosphine.

A7. P_2H_4

SUMMARY

The essence of this chapter is that binary compounds are named using
three systems. All the systems change the ending of the electronegative
element to -ide. The preferred systems, in addition, use prefixes or
Roman numerals (Stock method) to indicate the proportions of each
element. The use of prefixes is especially emphasized for binary compounds composed of nonmetals. The -ous/-ic method is not recommended. Certain trivial names for some hydrides have been approved.

The following questions are intended to give you further practice
using the techniques of this chapter. They will cover all aspects but in
no particular order.

Q1. Which of the following are binary compounds?
(a) O_2^{2-}
(b) NaI
(c) N_2O_3
(d) H_3PO_4
(e) Ca_3P_2

A1. b, c, and e

Q2. Write the formula for
(a) carbon disulfide
(b) ferric oxide

A2. (a) CS_2
(b) Fe_2O_3

Q3. Give the preferred name for
(a) P_2O_5 (b) Cl_2O_7

A3. (a) diphosphorus pentaoxide
(b) dichlorine heptaoxide

Q4. Write the formula for
(a) silane
(b) cupric fluoride (1+, 2+ common oxidation state for copper)
(c) magnesium bromide

A4. (a) SiH_4
(b) CuF_2
(c) $MgBr_2$

Q5. Name Cr_2O_3 according to all applicable methods. (Common chronium oxidation states are 2+, 3+, and 6+.)

A5. chromium(III) oxide, dichromium trioxide; some would state chromic oxide because they apply -ous to the lowest oxidation state and -ic to the next highest—this is to be discouraged.

Q6. State a preferred name for Hg_2Cl_2. What other names might also be given? (Common Hg oxidation states are 1+ and 2+.)

A6. mercury(I) chloride, dimercury dichloride, mercurous chloride, calomel (this is a trivial name that is frequently encountered in electrochemistry and medical terminology)

Q7. Name correctly:
(a) S_2Cl_2 (b) Na_4C (c) H_2S

A7. (a) disulfur dichloride
(b) sodium carbide
(c) hydrogen sulfide

Q8. Name correctly:
(a) HgI_2 (b) Li_3N (c) K_2Se

A8. (a) mercury(II) iodide or mercury diiodide
(b) lithium nitride
(c) potassium selenide

Q9. Name simple compounds composed of
(a) boron and nitrogen
(b) hydrogen and tellurium

A9. (a) boron nitride
(b) hydrogen telluride

Q10. Common oxidation states for mercury are 1+ and 2+. Using suffixes, write suitable names for the two compounds of mercury and bromine.

A10. mercurous bromide
mercuric bromide

Q11. Use the Stock system to name
(a) Al_2O_3 (b) CuH_2

A11. (a) aluminum(III) oxide
(b) copper(II) hydride

Q12. The following compounds are composed of nonmetals. Name them in the preferred way.
(a) S_2F_{10}
(b) N_2O
(c) P_3N_5

A12. (a) disulfur decafluoride
(b) dinitrogen oxide
(c) triphosphorus pentanitride

Q13. Write the formulas for
(a) manganese(IV) oxide
(b) vanadium(V) oxide

A13. (a) MnO_2
(b) V_2O_5

Q14. Give two acceptable names for AuF_3.

A14. gold(III) fluoride,
gold trifluoride;
auric fluoride will be seen
but not encouraged

Q15. Write the formula for ferrous oxide.
(Common oxidation states for
iron 2+ and 3+.)

A15. FeO

1. (2 pt)   What is a binary compound? Give an example.

2. (1 pt ea)   Give the preferred IUPAC name for each of the following:

   (a) CS_2

   (b) CaH_2

   (c) Al_2S_3

   (d) N_2O_4

   (e) NiO

   (f) PH_3

   (g) BaF_2

   (h) N_2H_4

3. (4 pt)   Give all reasonable names currently in use for $SnCl_4$. State the preferred name first. (The usual oxidation states are 2+, 4+.)

4. (3 pt)   Use the Stock system, if applicable, to name

   (a) V_2O_3

   (b) N_2O

   (c) $MnSe$

5. (2 pt)   Distinguish between As_2O_3 and As_2O_5 using suffixes in their names.

6. (5 pt)   Write the formula for each compound:

   (a) arsine

   (b) ammonia

   (c) ferric bromide

   (d) titanium(IV) sulfide

   (e) magnesium iodide

CHAPTER

7

IONS AND RADICALS

OBJECTIVES

The student should be able to

1. Define, with examples, each of the following: a cation, an anion, the prefixes per- and hypo-, the suffixes -ate and -ite, a radical (chemical type, of course), and a free radical.
2. Write from memory the preferred IUPAC name for any common monatomic ion if given its symbol, and vice versa.
3. Compose the preferred name for a polyatomic cation made by adding protons to a central atom.
4. Distinguish between and use properly oxonium ion and hydronium ion.
5. Write the common or trivial name for each of the following polyatomic anions: OH^-, O_2^{2-}, NH_2^-, CN^-, I_3^-, and N_3^-.
6. State the general principle used in naming polyatomic anions.
7. Compose names for anions derived from organic acids.
8. State and use the exception allowed when naming polyatomic anions made by adding oxygen to a central atom or ion.
9. Recognize by the name the presence of a radical in a compound.

1. (1 pt ea) Write a brief definition and give one example for each:
   (a) radical
   (b) hypo-
   (c) -ite

2. (1 pt ea) Write the preferred IUPAC name for each:
   (a) F^-
   (b) N^{3-}
   (c) Li^+
   (d) Se^{2-}
   (e) V^{3+} (can exist in other positive oxidation states)

3. (2 pt) To what ion and under what conditions should the name "oxonium" be applied?

4. (1 pt ea) State the trivial name for
   (a) OH^-
   (b) O_2^{2-}
   (c) SO_3^{2-}
   (d) ClO^-

5. (2 pt) What general principles are followed in naming a polyatomic anion?

6. (1 pt) The name for PON is phosphoryl nitride. What special type of species is present in this compound?

7. (2 pt) Write suitable trivial names that will distinguish among ClO_3^-, ClO_4^-, and ClO_2^-.

8. (1 pt) Pentanoic acid is $HC_5H_9O_2$. Name the anion derived from this acid.

INTRODUCTION

Ions and radicals are distinct entities that deserve separate attention in a nomenclature book. The naming of the simple ones is straightforward. The naming of complex ions gets very involved and most of that procedure is deferred to Chapter 10. (You can find definitions of cations and anions in Chapter 3.)

7-T-1

Monatomic cations should be named like the corresponding element except that an indication of the oxidation state should be given whenever there is more than one likely state. The word "ion" should be included in the name.

Examples: Na^+, the sodium ion; Fe^{2+}, the iron(II) ion; and Fe^{3+}, the iron(III) ion.

Q1. A cation has a _____ charge.

A1. positive

Q2. A monatomic cation contains only a single nucleus. true/false

A2. true

Q3. Give a recommended name for
(a) Li^+
(b) Mg^{2+}
(c) V^{3+} (also shows 4+, 5+)

A3. (a) the lithium ion
(b) the magnesium ion
(c) the vanadium(III) ion

Q4. Why wasn't it necessary to insert the Roman numerals in the names for Li^+ and Mg^{2+}?

A4. both ions have only one
usual oxidation state;
it would not be incorrect
to write "the lithium(I)
ion," but it wouldn't be
necessary

Q5. Give a suitable name for

(a) Ba^{2+}

(b) Cr^{3+} (also 2+, 6+)

(c) Fe^{2+} (also 3+)

A5. (a) the barium ion
(b) the chromium(III) ion
(c) the iron(II) ion or
the ferrous ion; the
latter would be accept-
able because iron has
just two usual states;
however, it would not
be preferred.

7-T-2

Cations that contain more than one atom, often called **polyatomic cations,** can be made in several ways. Those created by adding other ions or neutral atoms or molecules to a monatomic cation will be treated as complex ions and named according to rules given in chapter 10.

Polyatomic cations made by adding protons (hydrogen ions) to a monatomic anion are named by adding -onium to the root of the name of the anion element. Suppose that the P^{3-} anion combined with four protons to form PH_4^+. This polyatomic cation would be called the phosphonium ion. Other examples are AsH_4^+, the arsonium ion; H_3S^+, the sulfonium ion.

The H_3O^+ case is special. You can view this ion as an oxide ion with three attached protons or as a water molecule with one attached proton. Strictly speaking, H_3O^+ should be named the oxonium ion, and this usage is expected by the IUPAC whenever it is believed that the ratio of H^+ to water molecules is 1:1. The term "hydronium ion," which is also used for H_3O^+, is to be reserved for situations where the number of H^+ per water molecule is uncertain, such as is frequently true in aqueous solutions.

The polyatomic cation NH_4^+ is commonly called the ammonium ion rather than by its systematic name, nitronium ion. The trivial name is deeply embedded in chemical tradition and literature, and consequently is given special recognition.

Q1. A polyatomic cation is a multinuclear positive species. true/false

A1. true

Q2. A polyatomic ion usually contains just one kind of atom. true/false

A2. false

Q3. Polyatomic cations formed by the addition of H^+ to a negative ion use the suffix _____ added to the root of the positive/negative ion.

A3. -onium, negative

Q4. Write the preferred name for
(a) PH_4^+ (b) IH_2^+

A4. (a) the phosphonium ion
(b) the iodonium ion

Q5. Assuming direct evidence for a 1:1 ratio of H^+ to H_2O, which name is correct for H_3O^+: oxonium ion or hydronium ion?

A5. oxonium ion

Q6. In an aqueous solution of H^+ where the ratio of H^+ to H_2O is indefinite, one should refer to H_3O^+ as the _____ ion.

A6. hydronium

Q7. The ammonium ion has the formula
_____ .

A7. NH_4^+

7-T-3

The names for monatomic anions consist of the root of the name of the element and the ending -ide along with the word "ion." For example, H^- would be the hydride ion; F^-, the fluoride ion; Se^{2-}, the selenide ion; P^{3-}, the phosphide ion; and Si^{4-}, the silicide ion.

Q1. The ending of the element's name is modified when naming monatomic anions. true/false

A1. true

Q2. The word "ion" is an essential part of the name for any type of ion. true/false

A2. true

Q3. Give suitable names for
(a) S^{2-}
(b) C^{4-}
(c) N^{3-}

A3. (a) the sulfide ion
(b) the carbide ion
(c) the nitride ion

Q4. Is it necessary to indicate the oxidation state for monatomic anions?

A4. no

Generally, polyatomic anions formed by the addition of other ions or neutral atoms or molecules to a central atom or ion are named by attaching the ending -ate to the root of the name of the central atom or ion and then indicating the numbers of each added species with names and prefixes. Much more detail on the rules governing such names will be given in Chapter 10.

   Unhappily, there are two important types of exceptions to these rules that need to be discussed at this point. The first is simply that certain polyatomic anions are given the ending -ide, and unfortunately there is no easy way to determine which anions fall into this group. You must simply memorize them. The complete list is fairly long, but only the more common anions need be remembered by the beginning chemistry student. They are OH^-, the hydroxide ion; O_2^{2-}, the peroxide ion; NH_2^-, the amide ion; CN^-, the cyanide ion; I_3^-, the triiodide ion; and N_3^-, the azide ion.

Q1. Most polyatomic anions have names that end in -ide. true/false

A1. false

Q2. The more common ending for a poly-atomic anion is _____.

A2. -ate

Q3. The correct name for the NH_2^- ion is _____.

A3. the amide ion

Q4. Give preferred names for
(a) CN^- (b) OH^-

A4. (a) the cyanide ion
(b) the hydroxide ion

Q5. Compare the anion O^{2-} and O_2^{2-}. State the obvious difference and name each.

A5. in O_2^{2-} each oxygen atom has one negative charge, whereas the oxygen atom has two negative charges in O^{2-}:
O^{2-}, the oxide ion
O_2^{2-}, the peroxide ion

Q6. Two very similar-appearing yet very different anions are N_3^- and N^{3-}. Name each properly.

A6. N_3^-, the azide ion
N^{3-}, the nitride ion

Q7. The I_3^- ion is called _____.

A7. the triiodide ion

7-T-5

The other exception to the general rules for naming polyatomic anions occurs when the substance being added to the central atom is oxygen. Some typical anions of this type are SO_4^{2-}, BO_3^{3-}, SO_3^{2-}, and IO_3^-. Although the rules apply to oxygen perfectly well, it has long been customary to add prefixes and suffixes to the stem name of the central atom to indicate the presence and proportion of oxygen and to leave its name totally out of the final designation. Thus names such as sulfate ion, SO_4^{2-}; sulfite ion, SO_3^{2-}; hypochlorite ion, ClO^-; and perchlorate ion, ClO_4^-, will be found in your chemistry books.

The standard termination -ate is replaced by -ite when it is necessary to denote the next-lower oxidation state than the one in the standard. These endings are supplemented when needed by the use of prefixes: per- with -ate to indicate a yet-higher oxidation state, and hypo- with -ite (usually) to denote an even-lower oxidation state. These are nicely illustrated by the series of oxygen–chlorine anions listed in Table 7-1.

Table 7-1. Oxochlorine Anions

Anion	Name	Chlorine Oxidation State
ClO_4^-	Perchlorate ion	7+
ClO_3^-	Chlorate ion	5+
ClO_2^-	Chlorite ion	3+
ClO^-	Hypochlorite ion	1+

Q1. According to the IUPAC, polyatomic anions should have names that end in ——————— .

A1. -ate (review Section 7-T-4)

Q2. A prominent exception to the rules occurs when ——————— is the added species in the polyatomic anion.

A2. oxygen

Q3. The trivial names of these oxoanions do not contain the name of the element ——————— .

A3. oxygen

Q4. The presence and proportion of oxygen in oxoanions is indicated through the use of ——————— and ——————— .

A4. prefixes, suffixes

Q5. The termination -ate is replaced by ——————— when the next-lower oxidation state is to be denoted.

A5. -ite

Q6. The prefix _____ fre-
quently goes with -ite and denotes
an even _____
oxidation state.

A6. hypo-, lower

Q7. The anion ClO_4^- is named the perchlorate ion. What is the significance
of the prefix per-?

A7. per- used with -ate
means a higher oxidation
state than the -ate
state

7-T-6

The IUPAC has agreed to retain this system of trivial names for substances that have been known for a long time but expects us to use the
more systematic method (see Chapter 10) for lesser known anions and
newly discovered ones. As a beginning student you should learn the
names and formulas in Table 7-1 and be able to extend them by
analogy to the other members of the halogen family. In addition, the
following anions and their charges with their names need to be included
in your memory bank: NO_2^-, the nitrite ion; PHO_3^{2-}, the phosphite ion;
AsO_3^{3-}, the arsenite ion; and SO_3^{2-}, the sulfite ion.

Q1. Give the name or the formula for
(a) PHO_3^{2-} (b) the arsenite ion

A1. (a) the phosphite ion
(b) AsO_3^{3-}

Q2. Write the names for
(a) SO_3^{2-} (b) SO_4^{2-}

A2. (a) the sulfite ion
(b) the sulfate ion;
the higher oxidation state
in SO_4^{2-} uses the -ate
ending

Q3. Write formulas for
(a) the nitrate ion
(b) the nitrite ion

A3. (a) NO_3^-
(b) NO_2^-

Q4. The hypoiodite ion would be expected to have the formula

_____ .

A4. IO^-

Q5. IO_3^- has the trivial name of

_____ .

A5. the iodate ion

Q6. What is the name for AsO_4^{3-}?

A6. the arsenate ion

Q7. Starting with the highest oxidation state, list the bromine analogs for Table 7–1.

A7. BrO_4^-, the perbromate ion
BrO_3^-, the bromate ion
BrO_2^-, the bromite ion
BrO^-, the hypobromite ion

7-T-7

A class of anions that do not follow all the IUPAC nomenclature rules for complex ions but do end in -ate are those derived from organic acids. Generally, the names for these acids end in -ic. The corresponding anion's name is formed by replacing the -ic with -ate.

Examples: $HC_2H_3O_2$, acetic acid, gives the anion $C_2H_3O_2^-$, which is called the acetate ion; $HC_5H_9O_2$, pentanoic acid, gives $C_5H_9O_2^-$, the pentanoate ion; and $HC_7H_5O_2$, benzoic acid, gives $C_7H_5O_2^-$, the benzoate ion.

Q1. The names of organic acids end in

_____ .

A1. -ic

Q2. The name of the usual anion derived from an organic acid replaces -ic with _____ and drops acid and replaces it with

_____ .

A2. -ate, ion

Q3. $HC_3H_5O_2$ is propanoic acid. What is the name of $C_3H_5O_2^-$?

A3. the propanoate ion

Q4. If the oxalate ion is $C_2O_4^{2-}$, what would you predict for the formula of oxalic acid?

A4. $H_2C_2O_4$

Q5. Hexanoic acid is $HC_6H_{11}O_2$. Write the formula for its anion and name it.

A5. $C_6H_{11}O_2^-$, the hexanoate ion

7-T-8

A **radical** is an atom or group of atoms that contains at least one un-paired electron. Radicals are found in a number of compounds. A radi-cal is difficult to isolate in the free state but when you do, it is called a **free radical**.

Names of oxygen-containing (or other chalcogens) radicals are given special names ending in -yl: hydroxyl, OH; phosphoryl, PO; thiophosphoryl, PS; carbonyl, CO; uranyl(V), UO_2^+; and iodyl, IO_2^+. Radicals of this type are placed first in compounds and lead to names such as carbonyl chloride for $COCl_2$ and phosphoryl nitride for PON. You should be able to recognize the presence of the oxoradical from a name. Going the other way (i.e., deciding that a radical is a part of a substance) is very difficult by inspecting its formula. You will not normally be asked to do this.

> **Q1.** Radicals are atoms or groups of atoms that contain _____ .

A1. unpaired electrons

> **Q2.** A free radical is one that is found independent of a compound. true/false

A2. true

> **Q3.** The ending -yl signals an _____ .

A3. oxoradical

> **Q4.** Names such as nitrosyl sulfide should tell you that a _____ is present in the substance.

A4. radical

> **Q5.** The radical OH is called _____ .

A5. hydroxyl (note the difference between this and hydroxide OH^-)

SUMMARY

The following points were covered in this chapter: (1) the system of nomenclature for simple and polyatomic ions; (2) the two endings -ate and -ite, and the two prefixes, per- and hypo-; (3) the difference between the use of oxonium ion and hydronium ion; (4) the special character of oxoanion nomenclature; and (5) the nomenclature of radicals.

POST-TEST

1. (1 pt ea)   What do the following terms mean?
(a) free radical
(b) per-
(c) -ate when compared to -ite

2. (1 pt ea)   Give the trivial name for
(a) NO_2^-
(b) NH_2^-
(c) NH_4^+
(d) CN^-

3. (1 pt ea)   Write formulas for
(a) the triiodide ion
(b) the hypochlorite ion
(c) the carbide ion
(d) the calcium ion
(e) the chromium(III) ion

4. (½ pt ea)   Give the correct trivial names for
(a) SO_4^{2-} and SO_3^{2-}
(b) PHO_3^{2-} and PO_4^{3-}

5. (1 pt)   When is it acceptable to use the name hydronium ion to denote H_3O^+?

6. (1 pt)   Using formulas, differentiate between the oxide and peroxide ions.

7. (1 pt)   What ending signals the presence of a radical in a compound?

8. (2 pt)   Write trivial names to distinguish among IO_4^-, IO_3^-, and IO^-.

9. (1 pt)   Name the anion derived from butanoic acid.

8

ACIDS

OBJECTIVES

The student should be able to

1. Define an acid and a base.
2. Identify an acid from an examination of a formula.
3. Write the correct name for a common acid if given its formula, and vice versa.
4. Recognize and apply the system of suffixes and prefixes for naming acids.

1. (1 pt)   A very useful definition of acids states that they are
            _____ donors.

2. (2 pt)   Circle the acids in the following list: PH_3, H_3BO_4,
            C_2H_6, SiH_4, HF, H_2S, CaH_2.

3. (5 pt)   Write the common (trivial) names for
            (a) HBr (aqueous solution)
            (b) HNO_3
            (c) $HClO_4$
            (d) H_2CO_3
            (e) $H_2S_2O_3$

4. (2 pt)   Distinguish between H_3PO_4 and HPO_3 by using prefixes
            in their names.

5. (3 pt)   Write the formulas for
            (a) hypobromous acid
            (b) pyrophosphoric acid or diphosphoric acid
            (c) peroxonitric acid

6. (2 pt)   Write names for H_3AsO_4 and H_3AsO_3 that clearly dif-
            ferentiate between them.

INTRODUCTION

Names for acids have long enjoyed a well-established tradition. Attempts to reorganize and systematize this nomenclature are difficult because many well-known names would have to be changed. The nomenclature commission has therefore chosen rules that retain many older names but encourages systematic nomenclature for new acids.

Most of the acids that you will encounter in a beginning course fall into the older category, where the nomenclature depends heavily on prefixes and suffixes. Fortunately, you have already learned some of these, so naming acids will become a matter of applying what you already know plus a few new wrinkles.

8-T-1

For our purposes, an **acid** is defined (by Brønsted–Lowry) as a species that can donate hydrogen ions or protons. Generally, when dealing with acids, you assume an aqueous solution, although the foregoing definition of an acid does not make this limitation.

A **base** is defined (by Brønsted–Lowry) as a species that can accept protons.

The formula of an acid is usually written with the detachable hydrogen ions shown first; thus HBr, H_2SeO_4, and $HC_2H_3O_2$ should be recognized as acids, while $(CH_3)_2CO$ should not be so classified.

Q1. An acid is a substance that can donate

_____ .

A1. protons or hydrogen ions

Q2. When HI is dissolved in water, we find evidence for appreciable amounts of H^+ ions in the solution. The HI molecule can be described as an

_____ .

A2. acid

Q3. A base is a proton _____ .

A3. acceptor

Q4. The reversible reaction

$$NH_3(aq) + H^+(aq) \rightleftharpoons NH_4^+(aq)$$

is well known. Which species acts like a base?

A4. NH_3

Q5. If you examine the right-to-left reaction in Q4, you should predict that NH_4^+ is acting like an _____.

A5. acid

Q6. Circle those formulas that you would expect to behave like an acid: SiH_4, $HClO$, H_2CO_3, PH_3, C_2H_6.

A6. $HClO$, H_2CO_3

8-T-2

HBr and $HC_2H_3O_2$ are called **monoprotic acids** because they both can donate one H^+ per molecule under normal conditions. Using parallel reasoning, H_2SeO_4 is a **diprotic acid** and H_3PO_4 is a **triprotic acid**. Students are sometimes puzzled by the statement that $HC_2H_3O_2$ is monoprotic because it obviously contains four hydrogens. Experiments show that three of the hydrogens are more strongly bonded than the fourth, and further that these three do not dissociate under normal acid–base reaction conditions. This fact is indicated by showing them at a separate place in the formula. Remember, in general, that the first hydrogen(s) in the formula represent the detachable or acidic ones.

Q1. A triprotic acid would be expected to donate a maximum of _____ protons per molecule.

A1. three

Q2. Which of the following are diprotic acids? H_2SO_4, HBF_4, H_3BO_3, H_2S.

A2. H_2SO_4, H_2S

Q3. How many protons would you expect to be donated by the acid $H_2C_3H_2O_4$?

A3. two

Q4. How many protons can be donated by each of the following?
(a) $H_2C_2O_4$ (b) $HPHO_3$

A4. (a) 2
(b) 1

Q5. If you were asked to write a general formula to represent a diprotic acid, you would give _____ .

A5. H_2A, where A can be many different species

8-T-3

Acids that yield anions ending in -ide are named as binary compounds of hydrogen. Thus HBr would be called hydrogen bromide; HCN, hydrogen cyanide; H_2Te, hydrogen telluride; and so forth. Aqueous solutions of these substances are usually given different names by adding the prefix hydro- and the suffix -ic to the stem name for the electronegative atom and including the word "acid" in the final designation. An aqueous solution of HBr would be called hydrobromic acid; similarly, HCN in water would be hydrocyanic acid, and aqueous H_2Te would be named hydrotelluric acid.

Q1. Simple binary acids are named as compounds of _____ .

A1. hydrogen

Q2. Pseudobinary acids such as HCN are named as binary compounds of hydrogen. true/false

A2. true

Q3. The aqueous solution of a binary or pseudobinary acid is given a separate name. It is formed using the prefix _____ and the suffix _____ .

A3. hydro-, -ic

Q4. The name of the aqueous solution of HI is _____ .

A4. hydroiodic acid

Q5. Name the following:
(a) H_2Se (b) HF

A5. (a) hydrogen selenide
(b) hydrogen fluoride

Q6. What is the formula of the binary acid that gives hydroselenic acid?

A6. H_2Se

Q7. Whenever you encounter hydro-...-ic acid, you should immediately infer an aqueous solution. true/false

A7. true

Q8. Give the preferred name for HN_3.

A8. hydrogen azide

8-T-4

For acids containing more than two elements, the IUPAC preferred approach is to name them as hydrogen ...-ate, in accord with the general coordination principle to be seen in Chapter 10. Thus H_2SO_4 would become dihydrogen tetroxosulfate(VI); however, in deference to long-standing custom, (1) the endings -ous acid and -ic acid are permitted for the older acids, and (2) a number of common acids are allowed to keep their trivial names.

Common acids formed from anions ending in -ate are named ...-ic acid, where ... corresponds to the root or name of the central atom of the anion. From the sulfate anion we would get sulfuric acid. Similarly, the chlorate anion would give chloric acid and the perchlorate anion would yield perchloric acid.

If the name of the polyatomic anion ends in -ite, we name the corresponding compound a ...-ous acid. Thus the nitrite ion gives nitrous acid, the sulfite ion gives sulfurous acid, and the hypochlorite ion gives hypochlorous acid.

Q1. The IUPAC preferred method for naming acids formed from polyatomic anions is to use the word
_____ plus the suffix
_____.

A1. hydrogen, -ate

Q2. The preferred method is rarely used for common acids. true/false

A2. true

Q3. The endings -ous acid and
_____ are permitted for common acids.

A3. -ic acid

Q4. Give the common name for the acid formed from the SeO_4^{2-} ion.

A4. selenic acid

Q5. The NO_3^- ion would form an acid whose formula is _____ and whose trivial name is

_____ .

A5. HNO_3, nitric acid

Q6. The acetate ion has the formula $C_2H_3O_2^-$. Give the trivial name and corresponding formula of the acid derived from this anion.

A6. acetic acid, $HC_2H_3O_2$

Q7. Phosphoric acid is a relatively common substance. What do you predict its formula to be?

A7. H_3PO_4

Q8. State the trivial name for HBrO. (*Hint:* What is the name of the BrO^- ion?)

A8. hypobromous acid (derived from hypobromite ion)

Q9. Give the trivial name for H_2PHO_3 if PHO_3^{2-} is the phosphite ion.

A9. phosphorous acid

Q10. Give formulas for
(a) carbonic acid
(b) selenious acid

A10. (a) H_2CO_3 (from CO_3^{2-} anion)
(b) H_2SeO_3 (from SeO_3^{2-} anion)

8-T-5

As indicated in Section 8-T-4, a number of common oxo acids were allowed to keep their trivial names. Table 8–1 lists the acids you are likely to encounter in a first- or second-year course in chemistry. You should recall the prefixes per- and hypo- from Chapter 7. Their meanings are unchanged. Ortho-, meta-, pyro-, thio-, and peroxo- are all new. These will be examined in the next sections.

Table 8-1. Formulas and Trivial Names for Selected Oxo Acids

Formula	Trivial Name	Formula	Trivial Name
H_3BO_3	orthoboric acid or(mono)boric acid*	H_2SO_4	sulfuric acid
		H_2SO_5	peroxosulfuric acid
H_2CO_3	carbonic acid	$H_2S_2O_7$	disulfuric acid
H_4SiO_4	orthosilicic acid	$H_2S_2O_3$	thiosulfuric acid
$(H_2SO_3)n$	metasilicic acids	H_2SO_3	sulfurous acid
HNO_3	nitric acid	H_2CrO_4	chromic acid
HNO_4	peroxonitric acid	$HClO_4$	perchloric acid
HNO_2	nitrous acid	$HClO_3$	chloric acid
H_3PO_4	(ortho)phosphoric acid	$HClO_2$	chlorous acid
$(HPO_3)n$	metaphosphoric acids	$HClO$	hypochlorous acid
$H_4P_2O_7$	diphosphoric acid	$HBrO_3$	bromic acid
H_2PHO_3	phosphorous acid or phosphonic acid	$HBrO_2$	bromous acid
		$HBrO$	hypobromous acid
H_3AsO_4	arsenic acid	$HMnO_4$	permanganic acid
H_3AsO_3	arsenious acid		

*Parentheses indicate that the prefix is not commonly used.

8-T-6

Observe that H_4SiO_4 and H_2SiO_3 represent actual substances and that each has a silicon atom that shows an oxidation number of 4+. The difference between H_4SiO_4 and H_2SiO_3 is one water molecule. The nomenclature convention used to distinguish between these substances and others like them is to add ortho- to the trivial name for the species with the higher water content and to use meta- with the other. H_4SiO_4 becomes orthosilicic acid and H_2SiO_3 is a metasilicic acid.

> **Q1.** The prefixes ortho- and meta- are used to indicate the relative _____ content of two acids.

A1. water

Q2. H_3PO_4 and HPO_3 both contain a 5+ phosphorus atom. What trivial names should be assigned to distinguish between these two?

A2. H_3PO_4, orthophosphoric acid
HPO_3, metaphosphoric acid

Q3. If a certain metaboric acid has the formula HBO_2, what would you predict for the formula of orthoboric acid?

A3. H_3BO_3 (add one molecule of water to HBO_2)

Q4. An acid's name preceded by meta- can indicate that another acid exists that has greater water content. true/false

A4. true

Q5. The names for H_2SO_4 and H_2SO_3 do not contain the prefixes ortho- and meta-. Why?

A5. (1) the sulfur atoms have different oxidation states (2) they fail to meet the H_2O content criterion

8-T-7

Acids occasionally react with one another as follows:

$$H_3PO_4 + H_3PO_4 \rightarrow H_4P_2O_7 + H_2O$$

Note that two molecules of an acid have lost a molecule of water between them and formed a new acid. The new acid is named by adding di- to the trivial name of the parent acid; in the example, $H_4P_2O_7$ would be called diphosphoric acid.

Q1. When two molecules of an acid lose a water molecule between them, the resulting acid is named using the prefix ―――――――.

A1. di-

Q2. Predict the formula of the acid resulting from the loss of one H_2O between $H_2SO_3 + H_2SO_3 \rightarrow$ ―――――――.

A2. $H_2S_2O_5$

Q3. Give the common or trivial name for the product formed in Q2.

A3. disulfurous acid

Q4. What is the parent acid for $H_4P_2O_5$?

A4. phosphorous acid, H_2PHO_3

Q5. Give the common name for $H_4P_2O_5$.

A5. diphosphorous acid

Q6. Write the formula for disulfuric acid.

A6. $H_2S_2O_7$

8-T-8

When an oxygen atom in an oxo acid is replaced by a sulfur atom, a thio- acid is generated. When more than one oxygen is replaced, it is necessary to indicate the number of added sulfur atoms using a prefix. To illustrate, when one oxygen in carbonic acid (H_2CO_3) is replaced by a sulfur, the result is H_2CO_2S and the corresponding name is (mono)thiocarbonic acid. It is also possible to make H_2CS_3 trithiocarbonic acid and dithiosulfuric acid ($H_2S_3O_2$).

The prefixes seleno- and telluro- may be used exactly the same way for analogous acids.

Q1. A thio- acid is derived from an _____ acid.

A1. oxo

Q2. To form a thio- acid, you must replace an _____ atom with a _____ atom.

A2. oxygen, sulfur

Q3. To name most thio- acids, you should use two prefixes: _____ and _____ .

A3. thio-, numerical prefix

Q4. Write the formula for the (mono)thiosulfate anion.

A4. $S_2O_3^{2-}$ (from SO_4^{2-})

Q5. What would be the common name for the acid derived from $S_2O_3^{2-}$?

A5. (mono)thiosulfuric acid

Q6. The common name for H_3PO_3S is _____ .

A6. (mono)thiophosphoric acid

Q7. Name H_2SO_3Se.

A7. monoselenosulfuric acid

Once in a long while you will run into a peroxo . . . acid. These are derived from oxoacids by replacing an oxygen with —O—O—. HNO_4 is an example. It is made from nitric acid and is called peroxonitric acid. This name used to get shortened to pernitric acid, which created confusion for believers in the standard definition for per-. You might also see or hear of peroxosulfuric acid (H_2SO_5) in an early chemistry course.

SUMMARY

In this chapter you have seen how to name binary and pseudobinary acids and their aqueous solutions. You have also encountered the nomenclature for more complicated acids and have discovered that it depends heavily on prefixes and suffixes. It is essential that you know the meanings and uses of per-, hypo-, ortho-, meta-, thio-, di-, peroxo-, -ic, and -ous in this context.

Several review questions are listed below to give you further practice.

Q1. Using prefixes, differentiate between H_3BO_4 and HBO_3.

A1. orthoboric acid, H_3BO_4
metaboric acid, HBO_3

Q2. Name aqueous solutions of
(a) H_2Te (b) HF

A2. (a) hydrotelluric acid
(b) hydrofluoric acid

Q3. Name the acid made from
(a) IO^- (b) SeO_3^{2-}

A3. (a) hypoiodous acid
(b) selenious acid

Q4. Predict the formula for peroxosulfuric acid.

A4. H_2SO_5 (replace one oxygen with $-O-O-$)

Q5. What is the name of H_2PHO_3?

A5. phosphorous acid or phosphonic acid

Q6. When HCl is gaseous, it has one name; when it is in aqueous solution, it has another name. What are they?

A6. HCl (g), hydrogen chloride
HCl (aq), hydrochloric acid

Q7. Give the formulas for acids derived from the following anions:
(a) sulfate
(b) acetate
(c) peroxonitrate

A7. (a) H_2SO_4
(b) $HC_2H_3O_2$
(c) HNO_4

Q8. Name each acid in A7 and tell whether it is mono-, di- protic, and so on.

A8. (a) sulfuric acid, diprotic
(b) acetic acid, mono-
(c) peroxonitric acid, mono-

Q9. Name $H_2S_2O_3$.

A9. thiosulfuric acid

Q10. Write the formula for benzoic acid, given that the benzoate anion is $C_7H_5O_2^-$.

A10. $HC_7H_5O_2$

Q11. $H_2C_2O_4$ is the formula for oxalic acid. What two anions can be derived from it?

A11. $HC_2O_4^-$, $C_2O_4^{2-}$

Q12. Starting with orthophosphoric acid, how many different thio- acids could theoretically be made from it?

A12. four (replace each oxygen with sulfur)

Q13. Give the formula for each thio- acid for Q12 and name it.

A13. H_3PO_3S, monothiophosphoric acid
$H_3PO_2S_2$, dithiophosphoric acid
H_3POS_3, trithiophosphoric acid
H_3PS_4, tetrathiophosphoric acid

Q14. A di- acid is the result of losing a _____ molecule between two _____ molecules.

A14. water, acid

Q15. Predict the formula for disulfurous acid.

A15. $H_2S_2O_5$

Q16. Rhenium (Re) is an uncommon element that behaves much like manganese. Supply the formula for perrhenic acid.

A16. $HReO_4$

Q17. Technetium (Tc) belongs to the manganese family. What names would you predict for the following?

(a) H_2TcO_4 (b) $HTcO_4$

A17. (a) technetic acid
(b) pertechnetic acid

Q18. Using a prefix, distinguish between HNO_2 and $H_2N_2O_2$.

A18. HNO_2, nitrous acid
$H_2N_2O_2$, hyponitrous acid

1. (1 pt)   Using Brønsted–Lowry definitions, differentiate between an acid and a base.

2. (1 pt)   What distinguishing feature would you look for when scanning formula lists for acids?

3. (5 pt)   Match the formula with the correct name:

     (a) H_3PO_4 (1) thiophosphoric acid

     (b) H_2PHO_3 (2) diphosphoric acid

     (c) $H_4P_2O_7$ (3) phosphorous acid

     (d) HPO_3 (4) orthophosphoric acid

     (e) H_3PO_3S (5) metaphosphoric acid

                          (6) peroxophosphoric acid

4. (5 pt)   Give the common names for

     (a) H_3BO_3

     (b) $HBrO_3$

     (c) H_2SO_5

     (d) HNO_2

     (e) $HCN(aq)$

5. (3 pt)   State the prefixes, suffixes, or both that you would use to distinguish among each of the following pairs of acids.

     (a) $HClO_4$ and $HClO$

     (b) $H_2S_2O_7$ and $H_2S_2O_3$

     (c) H_3AsO_4 and H_3AsO_3

CHAPTER

9

SALTS

OBJECTIVES

The student should be able to
1. Name and write formulas from names, for binary and pseudo-binary salts.
2. Given the formula, compose a preferred name for salts containing polyatomic ions, and vice versa.
3. Recognize, name, and write formulas from names for acid salts.
4. Apply the rules for naming double and triple salts.
5. Recognize and name properly water of hydration.

1. (10 pt) Write the preferred name for each of the following.

      (a) CsI

      (b) $Ba(NO_3)_2$

      (c) $NH_4(ClO_3)$

      (d) $SrBr(OH)$

      (e) $Cu_2Cl(OH)_3$

      (f) $MgSO_4 \cdot 7H_2O$

      (g) K_2HPO_4

      (h) $RbNaCO_3$

      (i) $Na_6ClF(SO_4)_2$

      (j) $Ca_3(AsO_4)_2$

2. (10 pt) Supply formulas for each of the following.

      (a) calcium chloride

      (b) vanadium(IV) oxide sulfate

      (c) pentamagnesium fluoride (tris)phosphate

      (d) copper(II) tellurate pentahydrate

      (e) lithium dihydrogen arsenite

      (f) sodium peroxonitrate

      (g) barium hydroxide iodide

      (h) potassium hydrogen thiosulfite

      (i) aluminum sodium (bis)selenate

      (j) radium acetate dihydrate

INTRODUCTION

Salts should be thought of simply as compounds made up of positive and negative ions. Frequently, they can be traced directly to an acid–base reaction, yet many of them fit this relationship only indirectly. Salts that contain neither replaceable hydrogen ions or hydroxide groups are called normal salts,* while those that contain either H^+ or OH^- are acid salts and hydroxide salts. This chapter will give you the rules needed to name the given types of compounds.

9-T-1

Simple salts generally are binary or pseudobinary compounds and have already been discussed at length (see chapter 6). For review and practice purposes, some further examples are offered below:

> **Q1.** Give suitable names for
> (a) NaI   (b) $Ca(CN)_2$

A1. (a) sodium iodide
(b) calcium cyanide or
calcium dicyanide

> **Q2.** Name the following:
> (a) K_2O (b) CaF_2

A2. (a) potassium oxide or
dipotassium oxide
(b) calcium fluoride or
calcium difluoride

> **Q3.** Name each of the following
> (a) Rb_2Te (b) $MgCl_2$

A3. (a) rubidium telluride
(b) magnesium chloride or
magnesium dichloride

*William H. Nebergall, Frederic C. Schmidt, and Henry F. Holtzclaw, Jr., *General Chemistry* (Lexington, Mass.: D. C. Heath and Company, 1968), p. 319.

Q4. Write formulas for
(a) cesium bromide
(b) vanadium(III) chloride

A4. (a) CsBr
(b) VCl$_3$

Q5. Write formulas for
(a) chromium(III) oxide
(b) barium nitride

A5. (a) Cr$_2$O$_3$
(b) Ba$_3$N$_2$

Q6. Write three acceptable names for ZnF$_2$.

A6. zinc(II) fluoride
zinc difluoride
zinc fluoride

Q7. Write two acceptable names for CoS.
(The usual oxidation states for Co are
2+ and 3+.) Write the least preferred
name last.

A7. cobalt(II) sulfide
cobaltous sulfide

9-T-2

Normal salts containing polyatomic anions are named with the cation first followed by the anion. The previously learned conventions regarding prefixes and suffixes still hold.

Examples: Na$_2$SO$_4$, sodium sulfate; Ba$_3$(PO$_4$)$_2$, barium phosphate or barium orthophosphate; Ca(ClO$_4$)$_2$, calcium perchlorate; K$_2$SeO$_3$, potassium selenite; and Mg(ClO)$_2$, magnesium hypochlorite.

Note that whenever a polyatomic group appears more than once in a formula, it is enclosed by parentheses and its subscript placed outside.

Sometimes the multiplicative prefixes bis, tris, tetrakis, and pentakis help clarify when entire groups of atoms are indicated.

Q1. Normal salts do not contain replaceable _____ or _____ .

A1. hydrogen, hydroxide

Q2. Give suitable names for
(a) $Zn(ClO_4)_2$ (b) $CuBrO_3$

A2. (a) zinc (bis)perchlorate or zinc perchlorate
(b) copper(I) bromate

Q3. Supply names:
(a) K_3BO_3 (b) $Na_4P_2O_7$

A3. (a) potassium orthoborate or potassium borate
(b) sodium diphosphate

Q4. Give the formula for
(a) barium thiosulfite
(b) cobalt(III) arsenate

A4. (a) BaS_2O_2
(b) $CoAsO_4$

Q5. Write names for
(a) $Cr(NO_3)_3$ (b) $Mn(NO_2)_2$

A5. (a) chromium(III) nitrate
(b) manganese(II) nitrite

Q6. Supply the formulas for
(a) iron(III) sulfate
(b) mercury(II) nitrite

A6. (a) $Fe_2(SO_4)_3$
(b) $Hg(NO_2)_2$

Q7. Compose suitable names for
(a) K_2CrO_4 (b) VSO_4

A7. (a) potassium chromate
(b) vanadium(II) sulfate

Q8. Write the formulas:
(a) manganese(II) perchlorate
(b) nickel(II) carbonate

A8. (a) $Mn(ClO_4)_2$
(b) $NiCO_3$

Q9. The preferred names for Cu_2SO_4 and
$CuSO_4$ are _____
and _____ .

A9. copper(I) sulfate
copper(II) sulfate

Q10. The preferred names for these formulas are challenging:
(a) $Ni(NO_4)_2$ (b) $Mg(PH_2O_2)_2$
(c) $Rb_2S_2O_3$ (d) $SrMnO_4$

A10. (a) nickel(II) peroxonitrate
(b) magnesium hypophosphite
(c) rubidium thiosulfate
(d) strontium manganate

9-T-3

Some salts contain replaceable hydrogen as the result of partial neutralization of the parent acid. If H_3PO_4 is to be completely neutralized,
three molecules of the base MOH (M = any 1+ metal ion) are required:

$$H_3PO_4 + 3MOH \rightarrow M_3PO_4 + 3H_2O$$

Suppose that only one molecule of MOH per H_3PO_4 is available; then
the reaction would be

$$H_3PO_4 + MOH \rightarrow MH_2PO_4 + H_2O$$

The product MH_2PO_4 is known as an **acid salt** because it contains an anion with replaceable hydrogen, $H_2PO_4^-$. Diprotic and higher protic acids can participate in this type of partial neutralization; thus a fair number of acid salts are possible.

Compounds containing the hydride ion, H^-, are not to be considered acid salts.

Q1. An acid salt is distinguished from other salts by the presence of _____ hydrogen in the anionic portion.

A1. replaceable

Q2. Circle the acid salt(s)
$Ba(OH)Cl$ $KHCO_3$ LiH

A2. $KHCO_3$; look for anions that come from acids

Q3. List the possible acid salt anions that you would expect from sulfuric acid.

A3. HSO_4^-

Q4. Can an acid salt be derived from nitric acid?

A4. not under normal conditions

9-T-4

The names of acid salts are simply formed by inserting the word "hydrogen" and a numerical prefix, if needed, immediately in front of the name of the anion.

Examples: $LiHSO_4$ would be named lithium hydrogen sulfate; NaH_2PO_4, sodium dihydrogen phosphate; and $KHPHO_3$, potassium hydrogen phosphite or potassium hydrogen phosphonate.

It is important to note that the nonacidic hydrogen in the last case is included in phosphite or phosphonate and not with the other hydrogen.

Q1. To name an acid salt requires the use of a numerical prefix, if needed, and the word _____.

A1. hydrogen

Q2. Write suitable names for
(a) $NaHCO_3$ (b) $KHSO_3$

A2. (a) sodium hydrogen carbonate (sodium bicarbonate is common but not preferred, even though it is seen on baking soda boxes)
(b) potassium hydrogen sulfite

Q3. Name
(a) $CaHPO_4$ (b) LiH_2BO_3

A3. (a) calcium hydrogen phosphate
(b) lithium dihydrogen borate

Q4. Write the formulas for
(a) sodium hydrogen selenate
(b) calcium dihydrogen diphosphate

A4. (a) $NaHSeO_4$
(b) $CaH_2P_2O_7$

Q5. Name
(a) $Na_2H_2SiO_4$
(b) $MgHPO_5$

A5. (a) (di)sodium dihydrogen silicate
(b) magnesium hydrogen peroxo(mono)phosphate

So-called **double** and **triple salts** are not unusual. Some examples are $KNaCO_3$, $NaAl(SO_4)_2$, $Na_6 ClF(SO_4)_2$, $Mg(OH)Cl$, and LaFO. Each of these examples contains at least three different ionic species. To name salts like these, you need the following rules:

1. All cations precede anions both in name and formula.
2. Cations are listed in alphabetical order except for acidic hydrogen, which is usually handled in the name of the anion.
3. Anions with regular endings are cited in alphabetical order.
4. Anions that appear more than once are indicated by using the multiplicative prefixes bis (two), tris (three), tetrakis (four), and so on.

The names for the formulas given above as examples will be used to illustrate these rules: $KNaCO_3$ is potassium sodium carbonate, $AlNa(SO_4)_2$ is aluminum sodium (bis)sulfate, $Na_6 ClF(SO_4)_2$ would be (hexa)sodium chloride fluoride (bis)sulfate, $MgCl(OH)$ is magnesium chloride hydroxide, and LaFO is lanthanum fluoride oxide.

The parentheses in the names indicate that the prefixes may not be necessary.

Q1. Double and triple salts occur when _____ or more ionic species are together in a compound.

A1. three

Q2. Which of the following are either double or triple salts?
(a) $Cu_2(OH)_3 Cl$ (b) $Na_2 PHO_3$
(c) $KMgF_3$ (d) NaN_3
(e) CaH_2 (f) $Ca_5 F(PO_4)_3$

A2. (a) $Cu_2(OH)_3 Cl$
(c) $KMgF_3$
(f) $Ca_5 F(PO_4)_3$

Q3. The only criterion needed for listing the names of the anions in a double or triple salt is _____.

A3. alphabetization

Q4. Which comes first in the formula?
(a) Na^+ or NH_4^+
(b) SO_4^{2-} or PO_4^{3-}

A4. (a) NH_4^+ (alphabet)
(b) PO_4^{3-} (alphabet)

Q5. What is the multiplicative prefix for each of the following?
(a) four   (b) two

A5. (a) tetrakis-
(b) bis-

Q6. Compose a name for $NaNH_4HPO_4$.

A6. ammonium sodium hydrogen phosphate

Q7. What formula is needed for vanadium (IV) oxide sulfate?

A7. $VOSO_4$

Q8. Supply a name for $KMgF_3$.

A8. magnesium potassium fluoride

Q9. Name Na_2SnCl_6 as a double salt.

A9. (di)sodium tin(IV) (hexa)chloride

Q10. Give a name for $Ca_5 F(PO_4)_3$.

A10. (penta)calcium(mono)fluoride (tri)phosphate

Q11. Supply the formula of barium bromide hydroxide.

A11. BaBr(OH)

9-T-6

A common part of a formula for either simple or complicated substances is the **water of hydration**. This is shown at the end and is normally preceded by a dot. $MgNH_4PO_4 \cdot 6H_2O$ shows six waters of hydration. The water of hydration is included in the complete name by adding the correct numerical prefix and hydrate to the end of the name. The example above would be called ammonium magnesium phosphate hexahydrate. Sometimes the word "anhydrous" is placed in front of the compound's name to emphasize that no water of hydration is present.

Q1. Water of hydration is recognized by its position at the ———— of a formula.

A1. end

Q2. The complete name for such a compound will show at its———— a numerical prefix and the word ————.

A2. end, hydrate

Q3. Supply a name for $CuSO_4 \cdot 5H_2O$.

A3. copper sulfate pentahydrate

Q4. Write a formula for calcium chloride hexahydrate.

A4. $CaCl_2 \cdot 6H_2O$

Q5. Must the compound be either a double or a triple salt to have water of hydration?

A5. no

Q6. Compose a name for $NaZn(UO_2)_3(C_2H_3O_2)_9 \cdot 6H_2O$.

A6. sodium zinc triuranyl (nonakis)-acetate hexahydrate

SUMMARY

After completing this chapter, you should be able to differentiate among, name, and write formulas for normal, acid, hydroxide, double, triple, and even larger salts. A preferred ordering scheme is given for listing the names of the participating ions in these compounds which you need to know. Although the chapter is short, the number of compounds that fall under its rules is very large; thus you should study it carefully.

1. (1 pt)   Give the formula of an acid salt derived from selenic acid.

2. (2 pt)   Where in a formula do you usually see the water of hydration?

    What punctuation is used to indicate its presence?

3. (1 pt)   The following cations should be assumed to be part of a triple salt: Na^+, NH_4^+, and Mg^{2+}. Arrange them in the correct order for naming.

4. (1 pt)   Arrange the following anions in the order to be used for naming a double or triple salt: H^-, O^{2-}, SO_4^{2-}, OH^-.

5. (10 pt)   Supply the preferred name for each.

    (a) $As(OH)_2 Br$

    (b) $NH_4 NO_2$

    (c) $Be(HSO_3)_2$

    (d) $NaHCO_3$

    (e) $Ca_3 P_2$

    (f) $Na_6 BrCl(SeO_4)_2$

    (g) $VOCl$

    (h) $NaZn(UO_2)_3 (C_2 H_3 O_2)_9 \cdot 6H_2 O$

    (i) $Na_2 KHSiO_4$

    (j) $BaCl_2 \cdot 2H_2 O$

6. (5 pt)   Write correct formulas for

    (a) lithium potassium tellurite

    (b) strontium chloride

    (c) calcium chloride hypochlorite

    (d) sodium trihydrogen diphosphate

    (e) anhydrous copper(II) chloride

10

COORDINATION COMPOUNDS

OBJECTIVES

The student should be able to

1. Define, with examples, each of the following terms: complex ion, coordination compound, ligand, inner coordination sphere, polydentate, chelate, geometrical isomers, bridging ligand.

2. Name, using the IUPAC system, complex ions and coordination compounds if given their empirical or structural formulas.

3. Write the correct formula if given the systematic name for a complex ion or coordination compound.

1. (2 pt)   Distinguish between a metal complex ion and a coordination compound, and give an example of each.

2. (1 pt)   What is the relationship between a ligand and the metal ion in a complex ion?

3. (2 pt)   Differentiate between a chelate and bidentate ligand.

4. (1 pt)   What symbol is used to denote a bridging ligand?

5. (2 pt ea)   Name the following compounds:

(a)

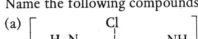
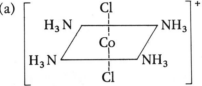

(b) $Na_3[Ag(S_2O_3)_2]$

(c) $[Cr(I)_2(PH_3)_4][Co(CN)_4(H_2O)_2]$ (*Hint:* Cr is 3+.)

(d) $[(NH_3)_5CO-O_2-CO(NH_3)_5]^{4+}$

(e) $[AlH_4]^-$

6. (2 pt ea)   Write the formulas (structural, if needed):

(a) *cis*-dibromodiammine platinum(II)

(b) tris(ethylenediamine)cobalt(III) sulfate (*Hint:* Ethylenediamine is bidentate and you may abbreviate it as en.)

INTRODUCTION

At first glance the names and formulas for the ions and compounds of this chapter look very formidable. Surprisingly, when approached systematically, they usually turn out to be relatively easy to name. Some students even enjoy putting these names together.

The application of the system of nomenclature of coordination compounds has proved to be very flexible and comprehensive. It can be used on a lion's share of inorganic compounds and could be applied to even more if other long-standing practices did not prevail.

The method is good enough to clarify many complicated cases, and a good rule to follow when in doubt is to try to name the substance using the principles of naming coordination compounds.

Several definitions need to be made and illustrated before the main thrust of naming and formula writing is taken up.

10-T-1

A **complex ion** consists of a central ion, or less frequently atom, which is attached to at least two other ions or molecules, called ligands.

$$\left[\begin{array}{c} Cl^- \\ \\ Cl^- \end{array} \underset{Pt^{2+}}{>} \!\!\!< \begin{array}{c} OH_2 \\ \\ Cl^- \end{array} \right]^-$$

Figure 1. Structural Formula for a Complex Ion.

The complex ion of Fig. 1 has a central ion, Pt^{2+}, three anionic Cl^- ligands, and one molecular ligand, H_2O. The charge of 1− is the net sum of the positive charges on Pt and negative charges on ligands. The lines between the center of coordination and the ligands represent coordinate covalent chemical bonds. The whole assembly is also called the inner **coordination sphere**. Complex ions may be positive or negative.

A neutral species containing a complex ion is called a **coordination compound**. These usually have rather complicated formulas with the complex ion enclosed in brackets.

Examples: $K_3[Co(NO_2)_6]$, $[Cr(H_2O)_4Cl_2]Br$, and $[Pt(NH_3)_2Cl_4]$.

Q1. The essential parts of a complex ion are a _____ ion or atom and surrounding _____.

A1. central, ligands

Q2. An inner coordination sphere is composed of a _____ and _____.

A2. central ion or atom, attached ligands

Q3. A complex may have a net charge of zero. true/false

A3. true

Q4. Ligands may be either _____ or _____.

A4. ions, molecules

Q5. Ligands are chemically bound to the center of coordination. true/false

A5. true

Q6. If $[Ni(NH_3)_6]^{2+}$ is a complex ion, which species are the ligands?

A6. NH_3 molecules

Q7. Coordination compounds contain at least one complex ion. true/false

A7. true

Q8. A complex ion behaves as a unit.
true/false

A8. true

Q9. The formula $[Pt(NH_3)_4Cl_2]Cl_2$ represents a coordination compound (with oxidation state Pt = 4+).
(a) Identify the complex ion by writing its formula.
(b) List the ligands in this complex.

A9. (a) $[Pt(NH_3)_4Cl_2]^{2+}$
(b) NH_3, Cl^-

Q10. Which of the following are coordination compounds?
(a) $[Ag(NH_3)_2]Cl$
(b) $[Ni(NH_3)_4]^{2+}$
(c) $K_2[PtCl_6]$
(d) $[Co(NH_3)_2Cl_2]$

A10. a, c, and d

10-T-2

A **ligand** is an ion or molecule capable of functioning as the donor partner in one or more coordinate covalent bonds. A coordinate covalent bond is a simple covalent bond where both the shared electrons originally were donated by one atom. The coordinating atom in a ligand is the one *directly* attached to the coordination center. When the ligand is bound to the center of coordination through only one atom, it is denoted as a **unidentate ligand**. The monatomic ions (e.g., Cl^-) must of necessity be unidentate, except in bridged species. Polyatomic ions and molecules are also often monodentate.

Some ligands may be multidentate; that is, they may contain more than one potential coordinating atom. These ligands are designated as **bidentate**, **terdentate**, and so on. The oxalate anion, $C_2O_4^{2-}$, from oxalic acid, $H_2C_2O_4$, can be considered to have two potential coordinating oxygen atoms and often acts as a bidentate ligand.

When a multidentate ligand is simultaneously attached by two or more donor sites to the same coordination center, forming a ring, the ligand is referred to as a **chelating ligand** and the resulting metal complex is often called a **metal chelate**. To continue with oxalate ion, it is called a chelating ligand when it participates in the following type of bonding arrangement:

$$O=C-O\diagdown$$
$$\quad\ \ |\quad\ \ \diagup M^{3+}$$
$$O=C-O\diagup$$

The resulting metal complex ion is the metal chelate. The ring structure is a characteristic of chelates.

Q1. A covalent bond is called coordinate when one atom donates _____ electrons to the shared pair.

A1. both

Q2. Species capable of donating electrons to a coordinate covalent bond function as _____ in coordination chemistry.

A2. ligands

Q3. Monatomic ions, because they may be attached at only one site, are called _____ ligands.

A3. monodentate

Q4. A terdentate ligand contains three _____ donor atoms.

A4. potential

Q5. A metal chelate is a metal complex ion that contains a chelate ligand.
true/false

A5. true

Q6. A multidentate ligand becomes a chelating ligand by bonding at two or more sites on the central ion in such a way as to form a _____ .

A6. ring

Q7. The organic substance 1,10-phenanthroline can be represented by

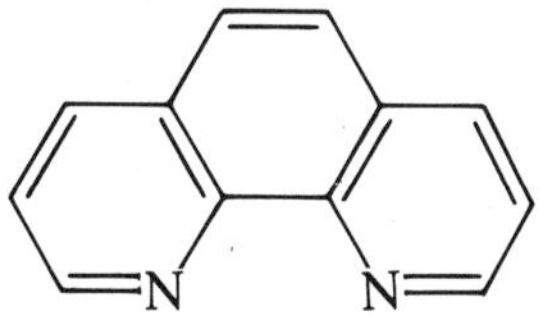

What ligand denticity would you predict for it?

A7. bidentate

10-T-3

A single coordinating atom in a ligand may attach simultaneously to two centers of coordination. In this role it is recognized as a **bridging** atom and its group as a bridging ligand. The following generalized structure illustrates this arrangement:

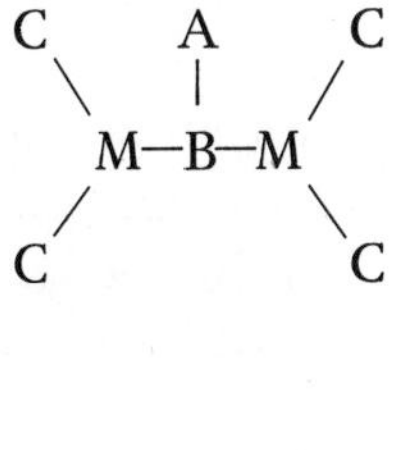

The A—B group is the bridging ligand and B would be the bridging atom. A bridging ligand is indicated by placing μ- before its name.

   Bridging can lead to **polycentric complexes**, that is, large complexes with two or more coordination centers held together by bridging ligands.

Q1. A bridging ligand contains a coordinating atom that is _____ bound to _____ centers of coordination.

A1. simultaneously, two

Q2.

$$\left[(en)_2 Co \overset{\overset{H_2}{N}}{\underset{\underset{H}{O}}{}} Co(en)_2 \right]^{3+}$$ is a

bridged complex. Specify the bridging ligands and the coordinating atoms.

A2. ligands: OH, NH
     atoms: O, N

Q3. The complex ion in Q2 is a binuclear complex. true/false

A3. true

10-T-4

Two species of the same composition but differing in structure are called **isomers. Geometrical isomers** are a major form of this complication and they are illustrated with the following example:

$$\begin{bmatrix} Cl & NH_3 \\ & Pt \\ H_3N & Cl \end{bmatrix} \quad and \quad \begin{bmatrix} Cl & Cl \\ & Pt \\ H_3N & NH_3 \end{bmatrix}$$

Both structures are square-planar. They have arrangements of ligands which are not identical. To demonstrate that the two arrangements are not actually identical, we superimpose the two structures and see if all like atoms coincide. No matter how you turn the structures, you will never be able to superimpose one exactly on the other.

Prefixes are used in front of the name of the compound to clearly identify each form. In the first structure, the like atoms are on opposite sides of the center of coordination. We indicate this "opposite" arrangement by the prefix *trans-*. The other form shows the like species on the same side of the center of coordination. The "same-side" arrangement is designated using *cis-*.

Q1. Isomers have different compositions and the same structures. true/false

A1. false

Q2. When two structures are not superimposable yet the compositions are identical, we have _____.

A2. isomers

Q3. The *cis-* form occurs when like species in the structure are _____ side(s) of the center of coordination.

A3. on the same

Q4. Which of the following pairs, assuming square-planar geometry, is the *trans-* form?

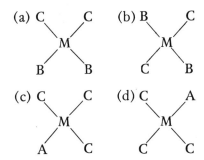

A4. (b) is *trans-* compared to (a);
(c) and (d) are not isomers;
you can superimpose them

This completes the necessary preliminary definitions; you are now ready to learn the IUPAC preferred method of naming complex ions and coordination compounds.

10-T-5

In complex ion formulas, except when obviously structural, the central atom or ion is placed first followed by the ligands in prescribed order (see the following sections). The whole complex assembly is usually enclosed in brackets and the net charge on the complex, if there is one, is shown as an upper-right superscript.

Examples: $[AlF_4]^-$, $[CoN_3(NH_3)_5]^{2+}$, $[Cr(SCN)_4(NH_3)_2]^-$, and $[Ag(S_2O_3)_2]^{3-}$.

In naming complex ions, the central atom or ion name is given after the ligands. The oxidation number of the central ion can be given after its name using Stock notation, or the charge on the complex can be given after the name using an Arabic number in parenthesis (Ewen–Bassett notation). Using the names for the examples above, you see, respectively, tetrafluoroaluminate(III) ion or tetrafluoroaluminate (1–) ion, pentaammineazidocobalt(III) ion or pentaammineazidocobalt (2+) ion, diamminetetrathiocyanatochromate(III) ion or diamminetetra-thiocyantochromate (1–) ion, and dithiosulfatoargentate(I) ion or dithiosulfatoargentate (3–) ion. Note that the Latin name for silver was used in the last name. Recall Section 4-T-1.

Q1. When naming a complex ion, the ligands follow/precede the central ion?

A1. precede

Q2. In formulas, the central ion precedes/ follows the ligands?

A2. precedes

Q3. The correct way of stating the oxidation number of the central ion in a name is 3+, 2+, and so on. true/false

A3. false

Q4. What is the correct way of showing the oxidation number of the central ion in a name?

A4. Stock notation

Q5. Criticize this representation of a complex ion: $[(NH_3)_6 Co]^{III}$. (Co is 3+.)

A5. (1) Co should precede NH_3's
(2) the charge on the complex is 3+
(3) don't use Stock notation here

Q6. The Ewens–Bassett notation gives the net charge on complex ion rather than the charge on the central metal ion. true/false

A6. true

Q7. Give the formula for hexaquachromium(III) ion. (*Hint:* Water in coordination spheres = aqua.)

A7. $[Cr(H_2O)_6]^{3+}$

10-T-6

The name of the central ion in complex cations and neutral complexes is unchanged when using it in the name for the complex; thus $[Pt(NH_3)_6]^{4+}$ is called hexammineplatinum(IV) ion. On the other hand, the name of the central ion in a complex anion should be given the ending -ate; thus $PtCl_6^{2-}$, following the rule, is named hexachloroplatinate(IV) ion.

The Latin root should be used in those cases mentioned in Section 4-T-1. To illustrate, $[AgCl_2]^-$ should be dichloroargentate(I) ion rather than dichlorosilverate(I) ion.

The only meaning for -ate in coordination chemistry is complex anion. The other endings that you know for anions do not apply here.

Q1. Complex ions that end in -ate have positive/negative charges.

A1. negative

Q2. A complex ion name that includes ...silver(I) ion must be referring to a

_____ .

A2. cation

Q3. Suppose that your classmate named a complex anion using the -ite ending. Why was this unnecessary and wrong?

A3. wrong because IUPAC decrees -ate as only anionic ending; unnecessary because Stock or Ewens–Bassett notation allows charges and oxidation state to be discovered

Q4. Which of the following should use Latin roots when acting as the central ion in complex anions?
(a) iron   (b) gold
(c) zinc   (d) copper
(e) tin   (f) lead

A4. all but zinc

10-T-7

The names for **cationic** and **neutral ligands** are simply the corresponding name without change for the cation or molecule. Thus, if PH_3 is a ligand, it would appear in the complex name as phosphine, and so on. Cationic ligands are very rare.

The major exceptions to this rule occur whenever water or ammonia behave as ligands. Under these circumstances they are named aqua and ammine, respectively. You should note carefully that certain nitrogen-containing organic molecules, such as RNH_2 and R_2NH, are called amines. These, too, can act as ligands, so the double-m spelling is very important if you mean NH_3.

Q1. A neutral ligand retains its name, without change, in a complex ion name. true/false

A1. true

Q2. A positively charged ligand would require a different ending than that for the cation name. true/false

A2. false

Q3. Water behaving as a ligand is called _____ . (Be sure you spell it correctly.)

A3. aqua

Q4. Ammine is for NH_3 when it behaves like a ligand. true/false

A4. true

Q5. It probably does not really matter if you misspell the ligand name for ammonia as amine because good chemists will know what you mean.

A5. wrong, amine represents an altogether different molecule

10-T-8

The names for **anionic ligands**, excepting those derived from hydro-carbons, end in -o.* To form the ligand name, the final e in each of the anion endings is replaced by -o, becoming -ido, -ito, and -ato, respectively.

Examples: SO_4^{2-}, sulfato; N^{3-}, nitrido; N_3^-, azido; NH_2^-, amido; and HSO_3^-, hydrogensulfito.

This name-forming rule is broken by the following -ide ions: F^-, Cl^-, Br^-, I^-, O^{2-}, O_2^{2-}, OH^-, H^-, and CN^-. Their ligand names are derived by replacing -ide with -o, so we get F^-, fluoro; Cl^-, chloro; O^{2-}, oxo; and so on. H^- is named hydrido or hydro and two other exceptions with special names are S^{2-}, thio; and SH^-, mercapto.

Anions derived from hydrocarbons are given radical names ending in -yl rather than -o. $C_6H_5^-$ becomes phenyl rather than phenido.

Q1. Almost all anionic ligands have names that end in _____.

A1. -o

*"Nomenclature of Inorganic Chemistry," *J. Am. Chem. Soc.*, **82**, 5538 (1960).

Q2. The exception to the statement above occurs with anions derived from

_____ .

A2. hydrocarbons

Q3. If an anion name ends in -ite, you form its ligand name by replacing this ending with -o. true/false

A3. false

Q4. What steps should you follow to properly form the ligand name for the anion in Q3?

A4. replace the e in -ite with o to get -ito

Q5. What are the corresponding anion formulas for each of the following?
(a) mercapto
(b) hydrogensulfato
(c) oxo
(d) phenyl
(e) hydro or hydrido

A5. (a) SH^-
(b) HSO_4^-
(c) O^{2-}
(d) $C_6H_5^-$
(e) H^-

Q6. Give ligand names for each of the following:
(a) I^-
(b) ClO^-
(c) OH^-
(d) NH_2^-
(e) S^{2-}

A6. (a) iodo
 (b) hypochlorito
 (c) hydroxo
 (d) amido
 (e) thio

Q7. What are the parent anions for each of the following?

 (a) methyl   (b) peroxo

A7. (a) CH_3^- (derived from CH_4 methane)
 (b) O_2^{2-}

Q8. Supply ligand names for

 (a) PHO_3^{2-} (b) SO_4^{2-}

A8. (a) phosphito
 (b) sulfato

10-T-9

When composing the name for a complex ion, the ligands are listed first, in alphabetical order, regardless of their charge.

Numerical prefixes used to specify the number of each ligand are not used in alphabetizing.

Q1. Which ligand is listed first in composing the name of a complex ion?

A1. use alphabetical order, ignoring numerical prefixes

Q2. Is water before/after ammonia in the order of listing?

A2. after; use aqua versus ammine

Q3. Arrange these anions in proper listing order: Cl^-, HCO_3^-, OH^-.

A3. Cl^-, HCO_3^-, OH^-

Q4. Arrange these ligands in proper listing order: NH_3, CO, S^{2-}, O^{2-}.

A4. NH_3, CO, O^{2-}, S^{2-}

10-T-10

The groundwork for naming coordination compounds has been laid and it is now time to put all the rules together. A list of steps follows that will help you approach these compounds in an organized fashion. An example, $[CoCl_2(NH_3)_4]Cl$, is worked out to illustrate each step.

1. *Determine if the species to be named is a complex ion or a co-ordination compound.*

$[CoCl_2(NH_3)_4]Cl$ is neutral; therefore, it is a coordination compound.

Q1. A coordination compound can be recognized by its _____ charge.

A1. lack of (neutral)

Q2. Which of the following are coordination compounds?
(a) $[Co(NH_3)_6]ClSO_4$
(b) $NH_4[Cr(SCN)_4(NH_3)_2]$
(c) $[CoN_3(NH_3)_5]^{2-}$
(d) $[Cu(C_5H_7O_2)_2]$
(e) $[Fe(en)_3][Fe(CO)_4]$

A2. a, b, d, and e

2. *Find the complex species in the formula.* Generally, they are enclosed by the brackets. Parentheses are normally reserved to set off the ligands.

$[CoCl_2(NH_3)_4]$ is the only complex species in the example.

3. *Determine the net charge on the complex ion(s).* Use charge balance (review Chapter 3).

$[CoCl_2(NH_3)_4]$ must be 1+, since it is combined with a single Cl^- ion to make the neutral compound.

Q1. Complex ions can be found in formulas by looking for the species enclosed by brackets. true/false

A1. true

Q2. Write the formulas for any complex ions in these compounds.
(a) $K_2[Pt(NO_2)_4]$
(b) $[Fe(en)_3][Fe(CO)_4]$
(c) $[Cr(H_2O)_6]Cl_3$

A2. (a) $[Pt(NO_2)_4]^{2-}$
(b) $[Fe(en)_3]^{2+}$ and $[Fe(CO)_4]^{2-}$
(c) $[Cr(H_2O)_6]^{3+}$

Q3. Identify the complex and give its charge in each case:
(a) $[CoCl(NH_3)_5]Cl_2$
(b) $K[Au(OH)_4]$
(c) $Na_3[Ag(S_2O_3)_3]$
(d) $[Cu(C_5H_7O_2)_2]$

A3. (a) $[CoCl(NH_3)_5]^{2+}$
(b) $[Au(OH)_4]^-$
(c) $[Ag(S_2O_3)_3]^{3-}$
(d) $[Cu(C_5H_7O_2)_2]^0$

4. *Determine the oxidation number of the central ion in the complex(es).* Once again use charge balance, only this time on all the species within the brackets. You will need to remember the oxidation numbers assigned to the common ligands.

$$(?) \quad + \quad (2-) \quad + \quad (0) \quad = 1+ \quad \text{so Co} = 3+$$
$$\text{for Co} \quad \text{for Cl}^- \quad \text{for NH}_3$$

Q1. What is the oxidation number for the metal ion in $[Al(OH)(H_2O)_5]^{2+}$?

A1. Al^{3+}

$$\underset{\text{for Al}}{(?)} + \underset{\text{for OH}^-}{(1-)} + \underset{\text{for H}_2\text{O}}{(0)} = 2+$$

Q2. What is the oxidation number for the metal ion in $[CrOF_4]^-$?

A2. Cr^{5+}

$$\underset{\text{for Cr}}{(?)} + \underset{\text{for O}^{2-}}{(2-)} + \underset{\text{for F}^-}{(4-)} = 1-$$

Q3. Give the oxidation number for the central metal ion in each complex:
(a) $K[Au(OH)_4]$
(b) $[Ru(HSO_3)_2(NH_3)_4]$
(c) $[CoN_3(NH_3)_5]SO_4$

A3. (a) Au^{3+}
(b) Ru^{2+}
(c) Co^{3+}

5. *Name the cation. If it is complex, use numerical (di, tri, tetra, etc.) and multiplicative prefixes (bis, tris, etc.) where needed, and name the ligands in alphabetical order.*

For our example, we should get tetraamminedichloro: don't break the name between ligands and don't drop the a in tetra even though it immediately precedes another vowel. Remember that you don't use the prefixes when alphabetizing the ligands.

Q1. Name the ligands $[Cr(SCN)_4(NH_3)_2]^-$ in the correct fashion.

A1. diamminetetracyanato

Q2. Supply the proper name for the ligands in $[Cr(O)_2O_2(CN)_2(NH_3)]^{2-}$.

A2. monoamminedicyano-dioxoperoxo

Q3. Write names for only the ligands in
(a) $[Co(NH_3)_6]ClSO_4$
(b) $K_3[Fe(CN)_5CO]$
(c) $[Fe(en)_3]$ (*Hint:* en is an abbreviation for ethylenediamine.)

A3. (a) hexaamine
(b) (mono)carbonylpenta-
cyano
(c) tris(ethylenediamine)

6. *Insert any prefixes indicating structural features, such as μ- or cis- or trans-.* You would have to have the structural formula to determine these; the empirical formula usually will not do.

Example: We cannot tell if these are present for the example with given information.

7. *Add the name of the central ion with proper ending and Stock or Ewen–Bassett notation to the already formed ligand name.* This is where knowledge of the charge on the complex and the oxidation state of the central ion is needed (steps 3 and 4).

The complex ion in the example is a cation, so there is no ending change for cobalt. The oxidation state for cobalt is 3+ and the net charge on the complex is 1+. Name: tetraamminedichlorocobalt(III) ion or tetra...cobalt(1+) ion.

Q1. The name of the central ion is not changed if it is part of a cationic complex. true/false

A1. true

Q2. What is the ending applied to a central metal ion that is part of an anionic complex?

A2. -ate

Q3. Do the endings -ite and -ide ever appear in complex naming?

A3. yes, but not preferred; the anion is sometimes given a trivial name from the older system

Q4. Give a complete name for the complex ion $[PtCl_3(NH_3)_3]^+$.

A4. triamminetrichloroplatinum(IV) ion
or tri...platinum(1+) ion

Q5. Name $[AgF_4]^-$.

A5. tetrafluoroargentate(III) ion

Q6. Supply names for
(a) $[Ru(HSO_3)_2(NH_3)_4]$
(b) $[Ag(S_2O_3)_2]^{3-}$
(c) $[Al(OH)(H_2O)_5]^{2+}$

A6. (a) tetraamminebis(hydrogensulfito)ruthenium(II)
(b) bis(thiosulfato)argentate(I) ion
(c) pentaaquahydroxoaluminum(III) ion

8. *Determine the name for the anion.* If it is complex, you will have to follow the prefixes, order of ligands, and endings just as before.

The anion in our example is simple: Cl^- or chloride.

9. *Combine the full name of the cation and the anion, in that order, to arrive at the name of the coordination compound.*

For the example, $[CoCl_2(NH_3)_4]Cl$ is named tetraamminedichlorocobalt(III) chloride or tetra...cobalt(1+) chloride.

Q1. Give a full name for $Li[AlH_4]$.

A1. lithium tetrahydroaluminate(III)
or lithium tetrahydroaluminate(1−)

Q2. Supply a name for
$[Co(NH_2)_2(NH_3)_4]Br$.

A2. diamidotetraamminecobalt-(III) bromide or di...cobalt(1+) bromide.

Q3. Name
(a) $K[AgF_4]$
(b) $Cs[ICl_4]$

A3. (a) potassium tetrafluoroargentate(III)
or potassium tetra...argentate(1-)
(b) cesium tetrachloroiodate(III)
or cesium tetra...iodate(1-);
iodine is 3+ in this complex

Q4. Name
(a) $Li[Sn(OH)_5H_2O]$
(b) $[Mn(CN)_5NO]^{3-}$

A4. (a) lithium(mono)aquapentahydroxostannate(IV)
or lithium...stannate(1-)
(b) pentacyanonitrosylmanganate(II) ion
or penta...manganate(3-) ion;
NO = neutral molecule, thus nitrosyl

Q5. Supply names for
  (a) $Cu_2^{II}[Fe(CN)_6]$
  (b) $[Fe(dipy)_3]Cl_2$ (dipy = dipyridine)
  (c) $[Co(en)_3]_2(SO_4)_3$ (en = ethylenediamine)

A5. (a) copper hexacyanoferrate(II); the small(II) superscript indicates oxidation state
  (b) tris(dipyridine)Iron(II)chloride or tris...iron(2+) chloride
  (c) tris(ethylenediamine)cobalt(III) sulfate or tris...cobalt(3+) sulfate

Q6. Compose a name for each:
  (a) $K_2[Cr(O)_2O_2(CN)_2(NH_3)]$
  (b) $H[Co(CO)_4]$

A6. (a) potassium(mono)amminedicyanodioxoperoxochromate(VI)
  (b) hydrogen tetracarbonylcobaltate(-I)

Q7. Name $[Fe(en)_3][Fe(CO)_4]$.
  (*Hint:* Fe in cation is 2+.)

A7. tris(ethylenediamine)iron(II) tetracarbonylferrate(-II)

Q8. Supply a name for $[Pt(py)_4][PtCl_4]$.
  (py = pyridine, Pt in the cation is 2+.)

A8. tetrapyridineplatinum(II)-tetrachloroplatinate(II) or tetra...platinum(2+)...platinate(2-)

Q9. Give the formula for tetraoxoferrate (2-) ion.

A9. FeO_4^{2-}

Q10. Write the formula for dichlorobis-(ethylenediamine)nickel(III).

A10. [NiCl$_2$(en)$_2$]

Q11. Supply the formula for tetrakis-(pyridine)platinum(II) ion.

A11. [Pt(py)$_4$]$^{2+}$

Q12. Give the formula for disodium penta-cyanonitrosylferrate(2-).

A12. Na$_2$[Fe(CN)$_5$NO]

Q13. What is the formula for potassium pentachloro(phenyl)antimonate(V)?

A13. K[Sb(Cl)$_5$(C$_6$H$_5$)]

Q14. Write formulas for
(a) sodium trioxosulfate(IV)
(b) sodium tetraoxosulfate(2-)

A14. (a) Na$_2$SO$_3$ (sodium sulfite)
(b) Na$_2$SO$_4$ (sodium sulfate)
(examples of new system encompassing the old)

Q15. Supply the formula for dipotassi-umtetranitrosyldithiodiferrate(I).

A15. K$_2$[Fe$_2$S$_2$(NO)$_4$]

10-T-11

Bridging groups* are specified by inserting μ-L-(L = name of ligand), di-μ-L- [or bis(μ-L-)], and so on. The name is separated from the rest of the complex's name by hyphens. Bridging groups are listed with

*W.C. Fernelius, K. Loening, and R.M. Adams, "Notes on Nomenclature," *J. Chem. Ed.*, **52**, 793 (1975).

the other ligands in alphabetical order. Whenever the same ligand is present both as a bridging and a nonbridging ligand, it is cited first as a bridging ligand. Several examples, which are also polynuclear complexes, will illustrate:

$$(NH_3)_5 CO-O_2-Co(NH_3)_5{}^{4+}$$

is named decaammine-μ-peroxo-dicobalt(4+) ion.

An example with two bridging groups:

$$\left[(NH_3)_4 Co \underset{O_2}{\overset{\underset{N}{\overset{H_2}{}}}{\diamondsuit}} Co(en)_2 \right]^{4+}$$

The name is μ-amido-tetraamminebis(ethylenediamine)-μ-peroxo-dicobalt(4+) ion. Another example:

$$PH_3 \quad Cl \quad Cl$$
$$Pd \quad Pd$$
$$Cl \quad Cl \quad PH_3$$

The name is di-μ-chloro-dichlorodiphosphinedipalladium(0).

Sometimes a bridging group bonds through two different atoms, such as $-NCS-$, whereas in another situation the same bridging group may provide the linkage using only one atom NCS<. When needed, these situations are indicated in the name by including the atomic symbols of the participating atoms; thus NCS< is thiocyanato-*S,S* and $-NCS-$ is thiocyanato-*N,S*.

Finally, a bridging group may be bound to more than two central atoms. A subscript following μ indicates the number of centers bound. For example, a suitable name for $[(CH_3 Hg)_4 S]^{2+}$ is tetramethyl-μ_4-thio-tetramercury(2+) ion. The subscript on μ tells us that four metal atoms are bound to the bridging sulfur atom.

Q1. Provide a name for

$$K_3 \left[\begin{array}{ccc} Cl & Cl & Cl \\ Cl-Tl-Cl-Tl-Cl \\ Cl & Cl & Cl \end{array} \right]$$

A1. potassium tri-μ-chloro-hexa-
chlorodithallium(3+)

Q2. Supply a name for

$$\left[(H_3N)_3\,Co \begin{array}{c} OH \\ \diagup \quad \diagdown \\ -OH- \\ \diagdown \quad \diagup \\ OH \end{array} Co(NH_3)_3 \right]^{3+}.$$

A2. hexaammine-tri-μ-hydroxo-
dicobalt(3+) ion

Q3. What is the structure of hexacarbonyl-
di-μ-iodo-dimanganese(2-) ion?

A3. $$\left[\begin{array}{c} CO \quad CO\ I\ CO \quad CO \\ \diagdown\diagup\ \diagup\diagdown\ \diagup \\ CO-Mn \quad Mn-CO \\ \diagup\ |\ \diagdown\diagup\ |\ \diagdown \\ CO \quad CO\ I\ CO\ CO \end{array} \right]^{2-}$$

Q4. What does the presence of μ-azido-
N,N in a name mean?

A4. azido is acting as a
bridging group using
only one N atom

(i.e., NNN<)

Q5. Suppose that you read μ-azido-*N,N'*.
What does this mean?

A5. azido is again acting as a
bridging group, but this time
two different nitrogen atoms
are participating (i.e.,
—NNN—)

Q6. μ_3 means that the bridging group is
bound to _____ metal
atoms.

A6. three

Q7. The most descriptive name for $[Cr_2Cl_9]^{3-}$ is tri-μ-chlorobis[tri-chlorochromium(III)] ion. What is its structure?

A7.

$$\left[\begin{array}{ccc} Cl & Cl & Cl \\ Cl-Cr-Cl-Cr-Cl \\ Cl & Cl & Cl \end{array} \right]^{3-}$$

10-T-12

Naming complex ions that are geometric isomers requires placing the prefix *cis-* or *trans-* at the very front of the systematic name. You need the structural formulas in order to name such complexes properly. The geometries that exhibit this property that you will encounter are square-planar and octahedral. Examples of each follow:

$$\left[\begin{array}{cc} PH_3 & Cl \\ & Pt \\ Cl & PH_3 \end{array} \right]$$

Square-planar
All atoms in the same plane with ligands at the corners, metal atom in the center.

trans-dichlorodiphosphineplatinum(II)
or *trans-* ...platinum(0)

The prefix *trans-* is used because like ligands are on opposite sides of the center of coordination.

$$\left[\begin{array}{cc} PH_3 & PH_3 \\ & Pt \\ Cl & Cl \end{array} \right]$$

cis-dichlorodiphosphine platinum(II)
or *cis-* ...platinum(0)

Like ligands are on the same side or adjacent to one another.

$$\left\{ \begin{array}{c} \text{NH}_3 \\ \text{H}_2\text{O} \diagup \mid \diagdown \text{OH}_2 \\ \text{Ni} \\ \text{H}_2\text{O} \diagup \mid \diagdown \text{OH}_2 \\ \text{NH}_3 \end{array} \right\}^{2+}$$

trans-diamminetetraaquanickel(II)
ion or *trans*-...nickel(2+) ion

Octahedral
Four ligands and central atom lie in same plane (like square-planar) and two ligands located directly above and below the central ion. All ligands are equidistant from the center of coordination. The solid figure formed by the faces has eight sides, thus is octahedral.

NH_3 groups are on opposite sides of the nickel.

$$\left\{ \begin{array}{c} \text{NH}_3 \\ \text{H}_2\text{O} \diagup \mid \diagdown \text{NH}_3 \\ \text{Ni} \\ \text{H}_2\text{O} \diagup \mid \diagdown \text{H}_2\text{O} \\ \text{H}_2\text{O} \end{array} \right\}^{2+}$$

cis-diamminetetraaquanickel(II) ion
or cis-...nickel(2+) ion

NH_3 groups are adjacent.

When looking at three-dimensional sketches of octahedral geometry, do not be fooled if the *cis*- or *trans*- ligands are not shown in the exact positions of the examples. Because all the ligand-to-metal ion distances are the same, several pairs of position for each form are equivalent.

Q1. Give the correct and complete name for

$$\begin{bmatrix} \text{Br} & & \text{NH}_3 \\ & \diagdown \diagup & \\ & \text{Ni} & \\ & \diagup \diagdown & \\ \text{H}_3\text{N} & & \text{Br} \end{bmatrix}$$

A1. *trans*-diamminedibromo-
nickel(II) or *trans*-...nickel(0)

Q2. Is the following structure *cis-* or *trans-*?

A2. *cis-*
(this is an example from
organic chemistry)

Q3. Name the following octahedral complex:

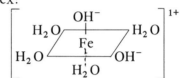

A3. *cis-*tetraaquadihydroxoiron-
(III) ion or *cis-*...iron(1+) ion

Q4. Draw the formula for *cis-*diiododi-
phosphinepalladium(II).

A4.

$$\begin{bmatrix} & I \quad I & \\ & Pd & \\ PH_3 & & PH_3 \end{bmatrix}$$

Q5. Name

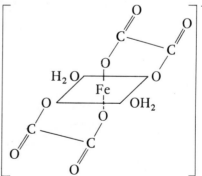

A5. *trans-*diaquadioxalatofer-
rate(III) ion or *trans-*...fer-
rate(1-) ion

Q6. Sketch *cis*-diaquadioxalatoferrate(1-) ion.

A6.

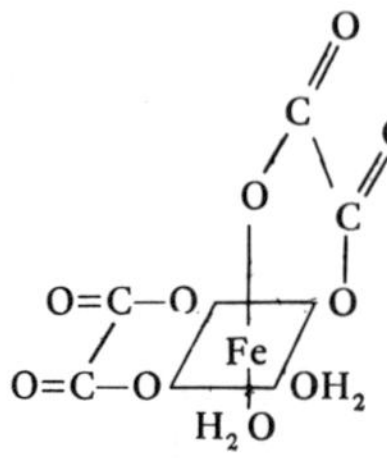

Q7. Draw *trans*-$[Ni(H_2O)_2(NH_3)_4]^{2+}$.

A7.

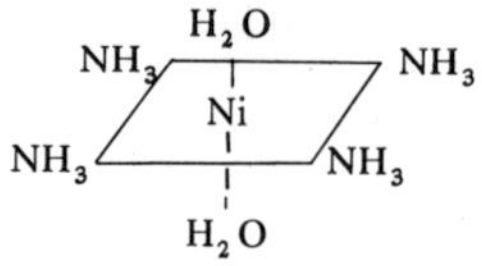

SUMMARY

This chapter defines ligand, inner coordination sphere, polydentate, chelate, geometrical isomers, and bridging ligand in the context of complex ions and coordination compounds. The rules for naming complex ions and coordination compounds are given, and an example worked out. Two special cases, geometrical isomers and bridging, are examined and examples of each are given to illustrate the naming of these species.

1. (2 pt)   Using an example, define a coordination compound.

2. (1 pt)   Assuming that the acetate anion behaves as a ligand, what would be its correct name in a complex ion?

3. (2 pt)   In each case, which comes first in the ligand ordering?

    (a) H_2O or NH_3 (b) OH^- or O^{2-}

    (c) N_3^- or NO_3^- (d) CO or NO^+

4. (1 pt)   What feature is especially important about a terdentate ligand?

5. (2 pt ea)   Name according to the IUPAC system:

    (a)

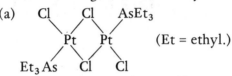

    (Et = ethyl.)

    (b) $K_2[Cr(O)_2O_2(CN)_2(NH_3)]$

    (c) $[Pt(py)_4][PtF_4]$

    (d) $[CoN_3(AsH_3)_3(PH_3)_2]SO_4$

    (e)

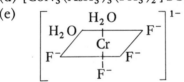

6. (2 pt ea)   Give the formulas for

    (a) carbonatobis(ethylenediamine)cobalt(III) selenate (Use en in place of formula of the diamine compound.)

    (b) *trans*-diamminedichloronickel(0)

I

KEYS TO PRE-TESTS

CHAPTER 1

1. (1 pt)   element
2. (2 pt)   When the elements are arranged in order of increasing atomic numbers, similarities in chemical and physical properties occur at regular intervals.
3. (2 pt)   The similarities of the chemical traits within families allow chemists to infer the chemical behavior of a large number of elements and compounds.
4. (1 pt ea)

Hydrogen

Alkali metals	Alkaline earths	Metals, all but those listed in right corner, and noble gases			C	N	O	F		
						P	S	U		
							Se	Br		
		Transition metals						I		
		*							Halogens	
		†								

*	Rare earths
†	Actinides

5. (½ pt ea)   (a) chalcogens: oxygen, O; sulfur, S; selenium, Se; tellurium, Te; polonium, Po

               (b) transition metals (first row): scandium, Sc; titanium, Ti; vanadium, V; chromium, Cr; manganese, Mn; iron, Fe; cobalt, Co; nickel, Ni; copper, Cu; zinc, Zn

               (c) shortest row: hydrogen, H; helium, He

               (d) alkali metals: lithium, Li; sodium, Na; potassium, K; rubidium, Rb; cesium, Cs, francium, Fr

               (e) alkaline earths: beryllium, Be; magnesium, Mg; calcium, Ca; strontium, Sr; barium, Ba; radium, Ra

6. (2 pt)   Chemical properties gradually change as you move across a row, whereas moving up or down a column produces little, if any, changes in chemical traits.

7. (½ pt ea)   (a) halogen

               (b) alkaline earth

               (c) chalcogen

               (e) alkali metal

CHAPTER 2

1. (1 pt)   The tendency of an atom to attract shared electrons is called electronegativity.

2. (2 pt)   (a) EN increases

            (b) EN increases

3. (1 pt ea)   (a) F,   (b) F,   (c) Li,   (d) N,   (e) C

4. (1 pt)   upper right corner

5. (1 pt)   lower left corner

CHAPTER 3

1. (1 pt)   1+

2. (1 pt)   2+, same

3. (2 pt)   2+

4. (1 pt)   4

5. (2 pt)   The combining power or valence is four, whereas carbon's apparent charge or oxidation number is zero.

6. (1 pt)   one

7. (1 pt)   1-

8. (1 pt)   1-

CHAPTER 4

1. (3 pt)   titanium, Ti; sodium, Na; argon, Ar; fluorine, F; chromium, Cr; selenium, Se; lead, Pb

2. (5 pt)   Cu, copper, cuprum; Sn, tin, stannum; Cl, chlorine; Li, lithium; Ag, silver, argentum; W, tungsten, wolfram

3. (2 pt)   This represents a diatomic, doubly positive ion of oxygen, mass number 18 for both atoms.

CHAPTER 5

1. (2 pt)   (a) CH_3

(b) C_2H_6

2. (1 pt)   (c)

3. (1 pt)   more electropositive

4. (1 pt) Ga_2O_3

5. (1 pt) $Ca_3(PO_4)_2$

6. (1 pt)   nitrogen

7. (2 pt)   Write the formula to correspond to the structure actually found experimentally.

acids and organic

8. (1 pt)   boron

CHAPTER 6

1. (2 pt) N_2O, NaBr

2. (9 pt)   (a) silver sulfide or disilver sulfide

(b) titantium triiodide

(c) diborane or diboron hexahydride

(d) zinc oxide

(d) aluminum selenide or dialuminum triselenide

(f) phosphorus pentabromide

(g) nitrogen dioxide

(h) iodine chloride

(i) ammonia

3. (2 pt) Cu_2O, copper(I) oxide; CuO, copper(II) oxide

4. (2 pt) VF_2, vanadium difluoride; VF_3, vanadium trifluoride

5. (2 pt) $FeCl_2$, ferrous chloride; $FeCl_3$, ferric chloride

6. (8 pt)   (a) Na_3N

(b) Ca_2C

(c) Co_2O_3

(d) MnS_2

(e) N_2O_4

(f) BaO

(g) SiH_4

(h) NH_3

CHAPTER 7

1. (1 pt ea)   (a) A radical is an atom or group of atoms that contains at least one unpaired electron: for example, OH, hydroxyl; CS, thiocarbonyl, UO_2^+, uranyl(V).

(b) hypo- means under or less than. In chemical usage it is generally taken to indicate an oxidation state less than that associated with -ous. If Cl is 3+ in $HClO_2$ (chlorous acid), then hypochlorous acid would be HClO, where Cl is 1+ or less than the -ous state.

(c) -ite is an ending used to indicate that the parent ion or compound was in an -ous oxidation state. The anion from HClO would be called the hypochlorite ion.

2. (1 pt ea)   (a) the fluoride ion

(b) the nitride ion

(c) the lithium ion

(d) the selenide ion

(e) the vanadium(III) ion

3. (2 pt)   Oxonium ion is strictly applied to H_3O^+ when it is known for certain that the ratio of H^+ to H_2O is exactly 1:1.

4. (1 pt ea)   (a) the hydroxide ion

(b) the peroxide ion

(c) the sulfite ion

(d) the hypochlorite ion

5. (2 pt)   Polyatomic anions are usually named by attaching the ending -ate to the root of the name of the central atom or ion and then indicating the numbers of each added species with names and prefixes.

6. (1 pt)   The ending -yl indicates the presence of a radical.

7. (2 pt) ClO_3^- is the chlorate ion

ClO_4^- is the perchlorate ion

ClO^- is the hypochlorite ion

8. (1 pt) $C_5H_9O_2^-$ is the pentanoate ion

CHAPTER 8

1. (1 pt) proton or hydrogen ion

2. (2 pt) H_3BO_4, HF, H_2S

3. (5 pt) (a) hydrobromic acid

(b) nitric acid

(c) perchloric acid

(d) carbonic acid

(e) thiosulfuric acid

4. (2 pt) H_3PO_4, orthophosphoric acid or phosphoric acid

HPO_3, metaphosphoric acid

5. (3 pt) (a) HBrO

(b) $H_4P_2O_7$

(c) HNO_4

6. (2 pt) H_3AsO_4, arsenic acid

H_3AsO_3, arsenious acid

CHAPTER 9

1. (10 pt) (a) cesium iodide

(b) barium nitrate or barium dinitrate

(c) ammonium chlorate

(d) strontium bromide hydroxide

(e) (di)copper chloride trihydroxide

(f) magnesium sulfate heptahydrate

(g) (di)potassium hydrogen phosphate

(h) rubidium sodium carbonate

(i) (hexa)sodium chloride fluoride (bis)sulfate

(j) calcium (di)arsenate

2. (10 pt) (a) $CaCl_2$

(b) $VOSO_4$

(c) $Mg_5F(PO_4)_3$

(d) $CuTeO_4 \cdot 5H_2O$

(e) LiH_2AsO_3

(f) $NaNO_4$

(g) $Ba(OH)I$

(h) KHS_2O_2

(i) $AlNa(SeO_4)_2$

(j) $Ra(C_2H_3O_2)_2 \cdot 2H_2O$

CHAPTER 10

1. (2 pt) A metal complex ion consists of a central ion, or less frequently an atom, which is attached to at least two other ions or molecules called ligands. A neutral species containing a complex ion is called a coordination compound. $Co(NO_2)_6^{3+}$ is a complex ion; $K_3[Co(NO_2)_6]$ is a coordination compound.

2. (1 pt) A ligand is an ion or molecule bound to the central metal ion in the complex.

3. (2 pt) A bidentate ligand has two potential binding sites. A chelate is a complex ion that contains a polydentate ligand.

4. (1 pt) The Greek letter μ denotes a bridging ligand.

5. (2 pt ea) (a) *trans*-tetraamminedichlorocobalt(III) ion or *trans*-...cobalt(3+) ion.

(b) sodium dithiosulfatoargentate(I) or sodium bis(thiosulfato) argentate(I)

(c) diiodotetraphosphinechromium(III) diaquatetracyanocobaltate(III)

(d) μ-peroxo-bis [pentaammine cobalt(III)] ion or pentaaminecobalt(III)-μ-peroxo-pentaaminecobalt(III) ion

(e) tetrahydroaluminate(III) ion (hydrido rather than hydro is acceptable)

6. (2 pt ea)

(a)

(b)

KEYS TO POST-TESTS

CHAPTER 1

1. (2 pt)   A period is characterized by a gradual but clear change in chemical and physical properties as you move across it, whereas a chemical family is distinguished by the close resemblance of chemical traits its members show to one another.

2. (½ pt ea)   (a) noble gas, (b) halogen, (c) alkaline earth, (d) chalcogen

3. (½ pt ea)   beryllium, Be; magnesium, Mg; calcium, Ca; strontium, Sr; barium, Ba; radium, Ra

4. (½ pt ea)   Cr, Zn, Ti

5. (2 pt)   When the elements are arranged in order of increasing atomic numbers, similarities in chemical and physical properties occur at regular intervals.

6. (2 pt)   The major benefit derived is the capability of inferring and predicting from the regularities in the periodic table a wide variety of chemical behavior.

7. (1 pt ea)

163

CHAPTER 2

1. (1 pt)   Electronegativity is a term that describes the tendency for an atom to attract shared electrons.
2. (1 pt ea)   (a) Li, (b) Mg, (c) F, (d) C, (e) Si
3. (1 pt)   fluorine, upper right corner
4. (1 pt)   EN increases
5. (1 pt)   EN decreases
6. (1 pt)   EN increases

CHAPTER 3

1. (1 pt)   1+
2. (2 pt)   Valence is the combining power of an atom in a compound; it is always positive. Oxidation number is the apparent charge of an atom in a compound; it may be either positive or negative.
3. (1 pt)   2-
4. (1 pt)   one
5. (1 pt)   2+
6. (1 pt)   1-
7. (1 pt)   6+
8. (2 pt)   2+; you normally expect 2-

CHAPTER 4

1. (3 pt)   zinc, manganese, chlorine, tin or stannum, gold or aurum, tungsten or wolfram
2. (3 pt)   silver, Ag; lead, Pb; sodium, Na; potassium, K; selenium, Se; chromium, Cr
3. (2 pt)   No; the charge should be 1+ or simply + and the atomic number should be in the lower position while the mass number (7) belongs in the upper position.
4. (2 pt) $^{2}\text{H}^{37}\text{Cl}$ represents a molecule of hydrogen chloride composed of a heavy isotope of both hydrogen and chlorine.

CHAPTER 5

1. (1 pt) Al_2S_3, both empirical and molecular
2. (2 pt)   empirical: CH_3; molecular: C_2H_6; structural:

```
    H   H
    |   |
H - C - C - H
    |   |
    H   H
```

3. (2 pt)   B, N, H, I, O, F
4. (1 pt)   CaO
5. (1 pt)   IBr
6. (1 pt)   The formula is written without regard for the actual structure or arrangement of the atoms.
7. (1 pt) $Ga_4(SiO_4)_3$
8. (1 pt)   No, because the anions CNO^- and NCO^- are not equivalent. They are two different substances, because of their structures.

CHAPTER 6

1. (2 pt)   A binary compound is composed of two elements. Examples are NO, Li_2S, KO_2, and $FeCl_3$.
2. (1 pt ea)   (a) carbon disulfide

                (b) calcium(II) hydride or calcium hydride

                (c) aluminum(III) sulfide or aluminum sulfide

                (d) dinitrogen tetraoxide

                (e) nickel(II) oxide

                (f) phosphorus trihydride or phosphine

                (g) barium(II) fluoride or barium fluoride

                (h) dinitrogen tetrahydride or hydrazine
3. (4 pt)   tin tetrachloride, tin(IV) chloride, stannic chloride
4. (3 pt)   (a) vanadium(III) oxide

                (b) not applicable

                (c) manganese(II) selenide
5. (2 pt) As_2O_3, arsenious oxide

                As_2O_5, arsenic oxide
6. (5 pt)   (a) AsH_3

                (b) NH_3

                (c) $FeBr_3$

                (d) TiS_2

                (e) MgI_2

CHAPTER 7

1. (1 pt ea)   (a) A free radical is an atom or group of atoms that contains at least one unpaired electron.

                (b) per- is used with -ate to indicate a higher oxidation state than that associated with -ate.

(c) -ate compared to -ite is used to indicate a higher oxidation state than that associated with -ite.

2. (4 pt) (a) NO_2^-, the nitrite ion; (b) the amide ion; (c) the ammonium ion; (d) the cyanide ion

3. (5 pt) (a) I_3^-, (b) ClO^-, (c) C^{4-}, (d) Ca^{2+}, (e) Cr^{3+}

4. (½ pt ea) (a) SO_4^{2-}, the sulfate ion; SO_3^{2-}, the sulfite ion;

(b) PHO_3^-, the phosphite ion; PO_4^{3-}, the phosphate ion

5. (1 pt) The hydronium ion may be used for H_3O^+ whenever the ratio of H^+ to H_2O is not known to be 1:1 (e.g., in many aqueous solutions).

6. (1 pt) O^{2-}, the oxide ion; O_2^{2-}, the peroxide ion

7. (1 pt) -yl

8. (2 pt) IO_4^-, the periodate ion; IO_3^-, the iodate ion; IO^-, the hypoiodite ion

9. (1 pt) the butanoate ion

CHAPTER 8

1. (1 pt) An acid is a proton donor and a base is a proton acceptor. The most common bases contain hydroxide ion as the proton acceptor.

2. (1 pt) Look for easily donated hydrogen ions (i.e., usually written first in the formula).

3. (5 pt) (a) H_3PO_4, orthophosphoric acid

(b) H_2PHO_3, phosphorus acid

(c) $H_4P_2O_7$, diphosphoric acid

(d) HPO_3, metaphosphoric acid

(e) H_3PO_3S, monothiophosphoric acid

4. (5 pt) (a) orthoboric acid or (mono)boric acid

(b) bromic acid

(c) peroxosulfuric acid

(d) nitrous acid

(e) hydrocyanic acid

5. (3 pt) (a) $HClO_4$, *per*chloric acid
  $HClO$, *hypo*chlor*ous* acid

(b) $H_2S_2O_7$, *di*sulfur*ic* acid
  $H_2S_2O_3$, *thio*sulfur*ic* acid

(c) H_3AsO_4, arsen*ic* acid
  H_3AsO_3, arseni*ous* acid

CHAPTER 9

1. (1 pt) $MHSeO_4$, where M = 1+
2. (2 pt)   At the end; the dot is used to indicate its presence.
3. (1 pt) NH_4^+, Mg^{2+}, Na^+
4. (1 pt) H^-, OH^-, O^{2-}, SO_4^{2-}
5. (10 pt) (a) arsenic bromide dihydroxide

   (b) ammonium nitrite

   (c) beryllium(bis)hydrogensulfite

   (d) sodium hydrogencarbonate

   (e) calcium phosphide

   (f) (hexa)sodium bromide chloride (bis)selenate

   (g) vanadium(III) chlorideoxide

   (h) sodium zinc triuranyl(nonakis)acetate hexahydrate

   (i) (di)sodium potassium hydrogensilicate

   (j) barium (di)chloride dihydrate

6. (5 pt)   (a) $LiKTeO_3$

   (b) $SrCl_2$

   (c) $CaCl(ClO)$

   (d) $NaH_3P_2O_7$

   (e) $CuCl_2$

CHAPTER 10

1. (2 pt)   A neutral species containing a complex ion is a coordination compound.
2. (1 pt)   acetato; acetate is not a hydrocarbon.
3. (2 pt)   (a) NH_3

   (b) OH^-

   (c) N_3^-; alphabetize

   (d) CO; neutral before cationic

4. (1 pt)   A terdentate ligand has three potential binding or donor atoms.
5. (2 pt ea) (a) di-μ-chloro-dichlorobis(triethylarsine)diplatinum(II) or chlorotriethylarsine platinum(II) di-μ-chloro-chlorotriethylarsineplatinum(II)

   (b) potassium(mono)amminedicyanodioxoperoxochromate(VI) or ...chromate(2+)

6. (2 pt ea)　　(a) $[Co(CO_3)(en)_2]_2\,SeO_4$

(b)

(c) tetra(pyridine)platinum(II)tetrafluoroplatinate(II)　　(tetrakis would also be acceptable prefix for pyridine)

(d) azidotriarsinediphosphinecobalt(III) sulfate [tetraoxosulfate-(VI) could be used for SO_4^{2-}]

(e) cis-diaquatetrafluorochromate(III) ion or cis-...chromate (2+) ion

BIBLIOGRAPHY

1. *How to Name an Inorganic Substance.* Oxford: Pergamon Press Ltd., 1977.

2. *Nomenclature of Inorganic Chemistry, Definitive Rules 1970*, 2nd ed. London: Butterworth & Co. Ltd., 1971.

3. "Nomenclature of Inorganic Chemistry," *Pure Appl. Chem.*, 28(1), 33 (1971).

4. W.C. Fernelius, K. Loening, and R.M. Adams, *J. Chem. Ed.*, 49, 488 (1972); 51, 468 (1974); 51, 603 (1974); 52, 793 (1975).

5. "Nomenclature of Inorganic Chemistry," *J. Am. Chem. Soc.*, 82, 5523 (1960).